Population and Environment Linkages

Population and Environment Linkages

Edited by

C.P. Prakasam
R.B. Bhagat

International Institute for Population Sciences, Mumbai

RAWAT PUBLICATIONS
Jaipur • New Delhi • Bangalore • Mumbai • Hyderabad • Guwahati

ISBN 81-316-0121-8

Published by
Prem Rawat for **Rawat Publications**
Satyam Apts, Sector 3, Jawahar Nagar, Jaipur 302 004 (India)
Phone: 0141 265 1748 / 7006 Fax: 0141 265 1748
E-mail: info@rawatbooks.com
Website: www.rawatbooks.com

New Delhi Office
4858/24, Ansari Road, Daryaganj, New Delhi 110 002
Phone: 011 2326 3290

Also at Bangalore, Mumbai, Hyderabad and Guwahati

Typeset by Rawat Computers, Jaipur
Printed at Chaman Enterprises, New Delhi

Contents

Foreword

The world today is passing through an environmental crisis of mammoth proportion. The land, water and air are increasingly polluted, causing severe strains on earth's precious life-support system. Although these days not many subscribe to the simple, Malthusian notion of population explosion as the main cause of environmental degradation, there is no doubt that the problem is man-made, by society's collective failure to internalize the environmental costs of modern, which becomes even more difficult to tackle from unabated population growth. It is also important to recognize that linkages between population and environment are bi-directional. Population size, growth, composition and distribution not only affect environment but also these attributes are being influenced by environmental degradation through its impact on mortality and migration. Cities are emerging as the main centers of environmental pollution with high density of vehicular traffic, industrial activity and growth of slum population. Health is another important area of mediation in the interface between population and environment. Even reproductive capacity may respond to environmental changes.

Keeping in view of the importance of population and environment nexus in contemporary research in social and demographic sciences, the Population-Environment Centre was established at IIPS in November 2004 by the Ministry of Environment and Forests, Government of India under the scheme for establishing a "Environment Information System (ENVIS)" network in the country to collect and disseminate information relating to environment. The POP-ENVIS Centre at IIPS was given the mandate of collecting and disseminating information related to population, human settlement & environment by adopting modern technologies of information acquisition, processing, storage and retrieval. To fulfill this objective, the center maintains the POP-ENVIS website, regularly brings out a POP-ENVIS bulletin and periodically organizes meetings and seminars.

The present volume is the outcome of selected papers presented in a seminar on 'Population Environment Nexus' organized by the POP-ENVIS Centre at IIPS, Mumbai on 21st October 2005. Population and environment linkages are complex and multifaceted, and no single volume can do justice to this burning issue of our times. But I am optimistic that the research papers included in this edited book would lead to a deeper understanding of the interface between water demand, sanitation needs, and growth of slums, household environmental factors and population dynamics. I deeply appreciate the efforts made by the coordinators and the team of professionals working in POP-ENVIS at IIPS in bringing out this book. I also take this opportunity to thank the support given by the Ministry of Environment & Forests, Government of India, in establishing and sustaining the Population-ENVIS Centre at our Institute.

15th August, 2006

P.N. Mari Bhat
Director
IIPS, Mumbai

Acknowledgement

Our sincere thanks are due to Dr. D. Bandyopadhyay, Director, Ministry of Environment and Forests, Government of India for providing grants for the National Seminar held on 21 October 2005 at Population ENVIS Centre, IIPS.

We would like to thank Ms. Reeta Khanna, Joint Director and Dr. S. Annadurai, National Programme Coordinator, Ministry of Environment and Forests, Government of India, New Delhi.

Our special thanks to Prof. P.N. Mari Bhat, Director, International Institute for Population Sciences and Major B.V. Ram Kumar, who supported us to organize the Seminar in the institute.

We appreciate the contributors for timely action and contribution of their valuable research papers.

Last, but not the least, thanks are due to Mrs. P. Ramani, Ms. Sonali Vadhavkar and the Web Designer, Mr. Anjani Kumar Mishra who really worked hard during the seminar, and assisted in bringing out this volume.

C.P. Prakasam
R.B. Bhagat

List of Contributors

Abdul Shaban, Tata Institute of Social Sciences, Deonar, Mumbai.

Alka Parikh, Reader, Department of Economics, University of Mumbai, Vidyanagri, Santacruz (E), Mumbai.

Anjali Radkar, Interdisciplinary School of Health Sciences, University of Pune, Pune.

Avishek Hazra, Research Scholar, International Institute for Population Sciences, Deonar, Mumbai.

Gopal K. Kadekodi, Director, Institute of Social and Economic Change, Nagarabhavi, Bangalore.

K. Radhakrishna Murty, Emeritus Fellow (UGC), Department of Sociology, Andhra University, Visakhapatnam.

K. Sita, 7, Beach Resort, Sector-10A, Vashi, Navi Mumbai.

Kamla Gupta, Professor and Head, Department of Migration and Urban Studies, International Institute for Population Sciences, Mumbai.

M. Ramakrishna Reddy, Obstetrician, Gynaecologist and Population Scientist, Srirajitha 160, 10th Main Road, 7th Sector, HSR Layout, Bangalore.

Malabika Ray, Vice Principal, SIES College of Commerce and Economics, Sion, Mumbai.

Mohua Guha, Research Scholar, International Institute for Population Sciences, Deonar, Mumbai.

N. Audinarayana, Professor and Head, Department of Population Studies, Bharathiar University, Coimbatore.

R.N. Sharma, Professor and Head, Unit for Urban Studies, Tata Institute of Social Sciences, Deonar, Mumbai.

Sancheeta Ghosh, Assistant Research Officer, Short Term Training Course Programme, International Institute for Population Sciences, Deonar, Mumbai.

Sandip Chakraborty, Research Scholar, International Institute for Population Sciences, Deonar, Mumbai.

Subhra Datta, Research Scholar, International Institute for Population Sciences, Deonar, Mumbai.

Sumati Kulkarni, Retd. Professor, E-2; Himachal Co-operative Housing Society, Plot No. 22, Sector-17, Vashi, Navi Mumbai.

T.V. Sekher, Reader, Department of Population Policies and Programmes, International Institute for Population Sciences, Deonar, Mumbai.

U.V. Somayajulu, Vice President, Head – Social Research, TNS India Pvt. Ltd., CICD Tower, Institutional Area, Hauz Khas, New Delhi.

Usha Ram, Reader, Department of Public Health and Mortality Studies, International Institute for Population Sciences, Deonar, Mumbai.

Overview

C.P. Prakasam and R.B. Bhagat

The present book comprises four sections pertaining to themes, namely, (i) population and environment linkages; (ii) food, water and sanitation (iii) urban environment and disaster; and (iv) household environment and health.

The seminar was inaugurated by Gopal K. Kadekodi, Director, Institute of Economic and Social Change, Bangalore. In his presentation entitled, "On Linkages between Population and Environment: Some Evidences from India", he showed how population growth, environmental degradation is linked with poverty at the state level analysis. The relationship between poverty and environmental degradation works in both directions. That is, an increase in poverty may impact negatively on the environment and the deterioration in the natural environment leads to increased poverty. However, there is no link found between population pressure and environmental degradation and poverty. According to Kadekodi, there is a lack of understanding of the role of institutions in mediating the linkages and the way in which institutions evolve in response to changes in the economic and policy environment. Apart from Kadekodi's presentation, three more papers dealt with the theme of population and environment linkages. Gupta and Guha in their paper entitled "Understanding the Linkages between Population Growth, Depletion of Natural Resources and its Implications for Women: A Focus on EAG States", tried to argue how environmental degradation creates additional burden for the womenfolk. Depletion of natural resources adversely affects the availability of fuel, fodder and water and the time required to obtain them. As such women's workload not only increases but their health deteriorates particularly among poor households if they lack to have adequate access to food. This paper culled data from various sources to support their argument.

The pressure of population on the limited natural resources in India is felt in several dimensions. The commonest sphere where such pressure is glaring is the already depleted or just disappearing common property asset in the rural landscape across the country. An observation of the pressure on the marginal land resources over the past one decade is presented by Reddy in his paper entitled "Vanishing Common Property Resources and their Environmental Impact in Rural Areas: Examples from Two Districts of Andhra Pradesh". Due to increasing population in the rural areas, the barren piece of land is being cultivated for food production. The scrub forests standing on the marginal lands have vanished. The denudation of scrub forest and removal of rocky boulders have led to the rain water run off, carrying with it the eroded top soil into water bodies consequent silting and reducing the capacity of the water bodies. The loss of grazing yards, reduced land fertility, and depleted water bodies in several parts of the selected districts in Andhra Pradesh.

In a paper by Sancheeta Ghosh entitled, "Concern for Environmental Degradation in India's Planning: A Review", highlights the trends in India's planning for the reduction of environmental degradation. Data have been used from the Planning Commission Reports of Government of India, from First Five-Year Plan up to the Tenth Five-Year Plan. The study focuses on two dimensions, such as outlays of plans in different domains of environment undertaken by the government and the total expenditure of the earmarked outlays. The study shows that there is an increasing importance given to the environment in planning and policies throughout the plan periods particularly from the Sixth Five-Year Plan.

In the study of the nexus between population and environment, food, water and sanitation are critical areas of concern, although interrelated. Several papers have dealt with these issues. Alka Parikh in her paper entitled, "Population vs. Food: Who is Winning the Race?", basically tries to show that the growth rate in food production has fallen during the 1990s after a spurt following the green revolution. This may be due to the fact that the capacity of the most fertile lands have been already reached and the less endowed regions need much more investment before they start giving a push to the food grain growth rates. She also looked into the future demand and supply of food grain in the light of changing nature of market forces.

There were three papers dealing with water from demographic, economic and ethical perspectives. Kulkarni in her paper entitled, "Water Resources in India: Issues, Concerns and the New Environment Policy", reviewed the availability of surface water in different river basins of the country, the present use of water in domestic, agricultural and industrial sectors of the growing economy of India. She also presented a critical analysis of various environmental policies dealing with the issue of availability and consumption of water by various sections of society. In similar vein, Radhakrishna Murty in his paper entitled, "Towards the Need for Water Ethics for Sustainable Development", refutes the notion that water is free and plentiful. He points out that today at least 400 million people of the world live in regions with severe water shortages. He further stated that we have been quick to assume rights to use water, but slow to recognize obligations to preserve and protect it. He quoted Thomas Fuller's saying, "we never know the worth of water till the well is dry". The paper further highlights the significance of and need for water ethic and its components as a part of sustainable development code that entails a wholly new approach to economic progress – one that harmonizes economic and population goals with ecological criteria. On the other hand, Shaban and Sharma present a study of water consumption in major cities of India. The paper entitled, "Water Consumption Patterns in Domestic Households in Major Indian Cities", is based on a survey in seven major cities namely Mumbai, Kolkata, Delhi, Hyderabad, Madurai, Kanpur and Ahmedabad. The paper also attempts to find out the extent of water deprivation in these cities by their socio-economic groups and areas. For effective management of water, the knowledge of water consumption pattern in various activities in domestic households is essential. The paper, therefore, examines the water consumption patterns activity-wise across socio-economic groups and their perception of water-wastage in various activities. The study shows that per capita water consumption, not due to lack of demand but due to lack of supply, is very low in major Indian cities. The per capita water availability in India is nowhere near the standards laid down by WHO or Bureau of Indian Standards (1993), and is also far below the other cities in the world, like Munich, Amsterdam, Sydney, Tokyo, etc. Water availability in the Indian cities varies widely with socio-economic groups and areas. Households with income below Rs 3,000 per month suffer a lot – about 72 per cent of such households in all the seven cities are water deficient. Among water consuming activities in households, the main water-guzzlers are: bathing and washing clothes. Household perception of water wastage is also highest in these two activities. Households in Kolkata waste more water – 51 per cent of water utilized is for bathing and washing clothes. The study recommends that not only the urgent step by the government is needed in augmenting as well as conserving the water resources in the cities, but also citizens have to play an important role in managing the water resources by cutting down over-consumption and wastages. Multipurpose use, recycling, and rainwater harvesting are some of the methods suggested for augmenting the declining availability of water resources in Indian cities.

The issue of sanitation is closely associated with environment and the use and disposal of water. Sekhar in his paper entitled, "Environmental Health and Sanitation Facilities: Challenges for India", attempts to analyse the availability of sanitation facilities both in rural and urban areas, and the strategies and experiences of various sanitation programmes undertaken in the country. He has used the available information from censuses, NSSO and other governmental and project sources. He has illustrated the experiences of some of the successful sanitation projects and highlighted the role of community participation and a sense of ownership by the users.

Three papers dealt with the issue of urbanization, environment and disaster. A paper on "Urban Ecological Issues with Special Reference to Mumbai" was presented by Sita and Ray. This paper was written on the backdrop of Mumbai deluge which occurred on 26 July, 2005. The authors argue that Mumbai flood 26/7 has brought the urban ecological issues into sharp focus. According to them, the ecological issues in urban areas can be broadly classified into three groups, namely, universal, regional and specific. Universal issues arise out of the intrinsic character of urban areas. Regional are those that are prevalent due to the characteristics of the area where the city is located. The specific are due to the distinctive features of the urban centre. Universal issues, found in all urban centres are primarily the result of the concentration of population and non-agricultural activities in these nodes. The land use pattern is modified and built-up areas and paved surfaces are predominant. The regional issues are a result of the characteristics of the macro and meso region in which the city is located. The specific issues are due to local relief, the site characteristics, the nature of a city's economy, etc. For example, Mumbai was initially a cluster of islands off the west coast of India. It acquired its present form due to reclamation over a long period of time. The suburban area has a central group of hills and is occupied to a large extent by the Sanjay Gandhi National Park. Streams rising in these hills and flowing to the sea were the drainage channels. One such important waterway is the Mithi River flowing into the Mahim creek. Extensive areas along the coast are low-lying and occupied by mangroves, marshes, mud flats, salt pans, etc. Unbridled urbanization has destroyed these ecologically fragile areas. The consequence was the recent flooding on an unprecedented scale that resulted in loss of life, immense human suffering and economic loss. In a similar paper, "Cyclones in Andhra Pradesh: Damages and Response" Somayajulu presented a brief history of cyclones in Andhra Pradesh and discusses the damage caused by these cyclones in terms of loss or property and human life and displacement, on the basis of secondary data and government documents. The paper also tried to present the findings of a study carried out among the households living in the coastal villages of the state in terms of cyclone preparedness, contingency plan and mitigation facilities, etc. The study indicates the need to improve the cyclone early warning mechanism and prepare cyclone contingency plan and provide mitigation facilities in the vulnerable villages.

A paper by Radkar entitled "Self-Reported Health Problems of Slum Dwellers in Pune", deals with the health status in segment of urban population that lives under unhealthy environmental conditions. The present paper makes an effort to understand the profile of slum population in Pune city and measures the morbidity level in one month reference period covering 1,100 households.

There were three papers on household environment and health status of children based on data derived from National Family Health Survey-2. The household environment is emerging as an important dimension influencing the health status of household population. The household environment is generally measured by the indicators like use of solid fuels, kachha and pucca houses, access to tap or covered water facility, source of lighting, provision of separate kitchen or cattle shed, etc. Thus according to Ram, Hazra and Chakraborty, household environment carries the biggest risk to the health of people at large and particularly to the poor, women and children. However, this risk is traditional in nature and is often associated with lack of development.

This paper discusses the existing scenario of household environment and health during infancy and childhood over a period of three decades in the major states of India and examines the association between health during infancy and childhood and a few household variables. In another paper, Audinarayana analysed the housing conditions and child morbidity particularly the incidence of diarrhoea, cough, and fever in the rural areas of two southern states namely Andhra Pradesh and Tamil Nadu. The studies shows that the prevalence of child morbidity conditions is much lower among those children living in households that have electricity as a major source of lighting, followed by households that have a separate kitchen and some type of toilet facility than their counterparts. Unprotected drinking water and higher room density have also showed somewhat negative effect on the prevalence of fever and cough. The last paper by Hazra, Datta and Guha analyses the effect of household environment on women's health in North-East India. The incidence of morbidities such as asthma, tuberculosis and jaundice are found to be higher in Meghalaya compared to other states of North-East India.

Although the papers incorporated in this volume reflects wide range of issues on the interface between population and environment, there are still many more issues left unaddressed. Migration and environment, one such area which needs more attention in future. Further, field based research in the areas of population and environment is lacking, as studies heavily draw upon the secondary sources which are often unable to throw light on the relationship between parameters of population and environmental conditions.

I Population and Environment Linkages

1 On Linkages between Population and Environment: Some Evidences from India

Gopal K. Kadekodi

The link between the state of environment and other socio-economic attributes is a complex one (Boulding, 1966; Markandya, 1998, Grossman and Krueger, 1995, 1996; Arrow et al., 1995; Seldon and Song, 1994; Shafik and Bandyopadhyay, 1992; Beckerman, 1992). It is highly site specific, be it at a watershed level, basin level, region, state or national level. Apart from population and its concomitants (age, sex composition, health status, etc.), other socio-economic components that are necessary to be linked and analysed are: income level and its degree of inequality, human development and poverty, and property rights on resources (Kadekodi and Venkatachalam, 2005). A clear understanding of the causality, implication, feedbacks and complexity is possible only with long time series data and well designed econometric models (Kadekodi, 2001; Duraiappah, 1998).[1]

As far as environment and natural resources are concerned, both the 'stock' of them and the 'flow' of environmental functions and services, income and welfare from the stock are to be evaluated to understand the links between human and environmental development. Human dimensions are also to be viewed as stocks (e.g., population) and flows (birth and death rates, mortality and morbidity rate and so on). It is the interplay of these human and natural resource endowments that bring about the changes in both and ultimately affecting sustainable development.[2]

The population of India is currently rising on an average at 2.15 as a decadal average, which has shown a decline over the recent decades. However, major states like Bihar, Haryana, Jammu & Kashmir,

Rajasthan and Uttar Pradesh are showing quite high growth rates, much above the national average. Moreover, Kerala, Bihar, West Bengal, Uttar Pradesh, Tamil Nadu and Punjab continue to be high on population density, to which several new states like Haryana and Assam have joined in 2001. During the last two decades, states like Andhra Pradesh, Bihar, Gujarat, Haryana, Maharashtra, Punjab and West Bengal have registered more that 50 per cent increase in the population densities. The situation is similar in terms of degree of urbanization as well.

India is endowed with 675,540 sq kms of forest cover (i.e., 20.55% of the total geographical area of 3,287,263 sq kms), of which about 416,809 sq kms are under dense forest cover. The country is also having about 7,515 kms of coastal length, rich in fishery and marine resources. About 84,148 sq kms of the geographical area of the country is under various rivers and streams, with an additional area of about 21,960 sq kms under lakes, reservoirs and canals. In terms of biodiversity, India is rich with two major hotspots (out of eight in the world), over 45,000 plant species and 81,000 animal species. However, with the growth of industrialization and population, apart from scarcity and threat to livelihood supports, a large number of environmental problems have cropped up, among which the air and water pollutions dominate, followed by degradation of biodiversity. Against this background, the amount of wastelands has been estimated currently to be 638,518 sq kms (or about 19% of the geographical area of the country).

What are the major changes taking place in the arena environment? According to the estimates made by the Planning Commission, between 1987 and 1997, states like Andhra Pradesh have witnessed highest deforestation of over 14 per cent, the least loss being in Kerala. Apart from traditionally known rivers such as Saraswati having vanished, both the groundwater tables and catchments of the river basins have lost their capacities as 'catchments'. As documented by Agarwal and Narain (1997), traditional water harvesting structures such as tanks have significantly vanished in many parts of the country. As a source of irrigation, between 1970-71 and 1997-98 the area under tank irrigation has come down from 4.1 to 3.1 million hectares (Kadekodi, 2004). Because of their concentration in the industrial and heavy transport intensive areas, the emission loads from these sectors cannot be ignored, though, on a per capita or per geographical area basis they may sound too small. As of 1990, about 2.186 million tonnes of carbon monoxide is being emitted annually from various industrial and transport sectors in India. Likewise, in 1990, about 156 million tonnes of carbon emission took place from the Indian continent, which rose to 251 million tonnes by 2001.[3] Such massive emissions of pollutants are acting against the health and quality of life. The forest degradation, depletion of water bodies, and degradation of ground and surface water qualities, deterioration of soil qualities due to excessive use of chemical fertilizers and pesticides, non-disposal or non-recycling of urban and other solid wastes, climate change, etc., are all increasing. They are resulting in incidences of adverse effects on human and animal lives.

The deterioration of health status, increasing incidences and re-emergence of diseases such as malaria, and other parasitic attacks due to climate change, etc., are coming in the way of human development. According to WHO, about one million people die of malaria annually in the world today. Increased incidences of respiratory diseases and heart disorders are attributed to urban and industrial air pollution. According to the estimates made by the World Health Organization (WHO), about 3.3 million people die every year from diarrheal diseases globally and at any one time there are 1.5 million with parasitic worm infections stemming from human excreta and solid waste in the environment (Bojö et al., 2001).

This paper is aimed at highlighting the possible linkages between environmental changes with the dynamics of population and the complexities involved in establishing or testing the linkages or adversaries between them and human development.

International View on Environment, Population and Development

Three different types of links between environmental matters and their changes are affecting the quality of life. They are the increasing blue environmental matters (such as pollution and degradation of air and water quality), second, depletion and deterioration of resources (such as forest, land and water bodies), and third, loss of links between livelihood supports to population and natural resources. Apart from the environmental status, population growth and its dynamics (migration, fertility rate, etc.) do also influence the state of development. On a global basis there is some evidence by now to say that environmental degradation affects the quality life, as much as changing quality of life in turn affects (or interferes with) environmental quality and status (Dasgupta, 2003; Duraiappah et al., 2000; Lopez, 1997; Markandya, 2001). Essentially, there is an increasing realization that the process of growth is making increasing demands on natural and environmental resources. This in turn is making adverse impacts on the environment, which in turn is affecting the human development itself. Figure 1 represents such an understanding of this feedback process of pressures and impacts. Income, levels of human development, extent of poverty and inequality are all determined by both the population dynamics and the status of environment. They in turn influence the state of environment.

It is in the context of the debate on the inverted U-shaped relation between income inequality, poverty and environment, that the two way linkages between poverty-environment-poverty is important but very difficult to establish (Dasgupta, 2003; Duraiappah, 1998; Dasgupta and Maler, 1994). How poor people are often impoverished by degradation in environmental resource base, which in turn forces them to degrade the environment further was the argument of the 'vicious circle' of the human development-environment nexus

Figure 1

Links between HDI, GDP, Population, Poverty and Environment

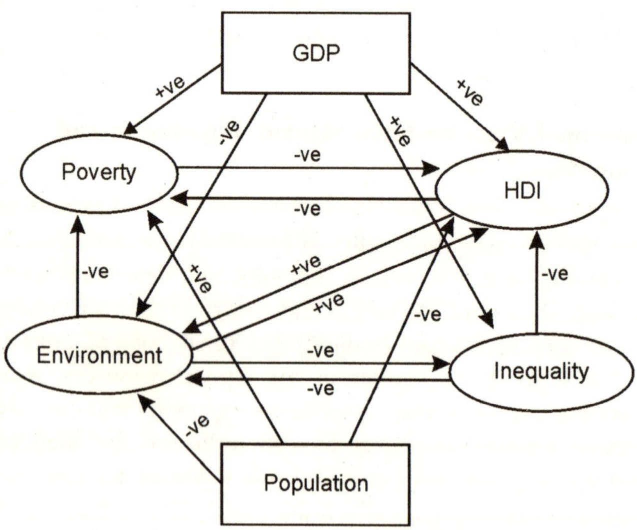

(WCED, 1987). For instance, environmental degradation is found to be mediated by high fertility rate, and in many cases poverty, population growth and environmental degradation feed up one another. This hypothesis has been criticized with a view on the failure of the society and the country to internalize the positive livelihood linkages and strengths of the link between them, and also for not being able to discourage the adverse pressures from development on environment through proper policy interventions (Nadkarni, 2001; Markandya, 1998; Jodha, 1986, 2000). After all, the increasing industrial and transport related air and water pollutions are not the result of the poor accessing to resources. But the policies on control of such environmental degradations are not properly addressed to make the polluters responsible for such polluting activities leading to degradations. That is why, it is often said to be a matter of proper institutions to link the human and resource development together. It is also evident from the fact that the governments all over the world are not able to introduce effective economic instruments (e.g., fiscal and monetary measures) to drive the use of natural resources towards their sustainable use, and also to put a cap on pollutions (Datt et al., 2004).

Based on several studies on the nexus between environment, population and poverty the following conclusions can be drawn:

- The causal relationship between poverty and environmental works in both directions. That is, an increase in poverty may impact negatively on the environment and the deterioration in the natural environment leads to increased hardship.
- Some of the relationships based on the evidences can be conflicting. For example, there is no simple relationship between increase in poverty and economic growth, the causal structure is quite complex. Nor is there an accepted relationship between the quality of the environment and changes in population;
- The linkages between poverty and the environment often work through changes in GDP, population, etc. The dynamics of the relations are only partly understood; and
- There is lack of understanding of the role of institutions in mediating the linkages, and the way in which institutions evolve in response to changes in the economic and policy environment.
- Population stress does not seem to have any significant direct effect on all aspects of environmental status in India.

Status at the National Level

The context being highly populated country like India, the debatable issue is whether population pressure alone drives the environmental degradation and hence in turn bring about the down turn of human development or, is it development which drives environmental degradation in the early phases of development as an inevitable consequence, a hypothesis referred as inverted U curve on environmental degradation (or popularly stated as Environmental Kuznets hypothesis).

Talking about population pressure, it may be useful to take a look at the recently estimated Environmental Sustainability Indices (ESI) for India (and likewise for 145 other countries in the world) by the Yale Centre for Environmental Law and Policy (2004), based on 76 variables.[4] It is found that the major variables and factors that determine the environmental status in India are air, water, land and biodiversity quality degradation. Among the policy variables, reduction in waste and air pollution, changing consumption patterns are important. Equally important are reducing natural disasters and water stress. Then comes the policies on private sectors initiatives and international responsibilities. Policy on reduction in population pressure (viewed by population growth rate and total fertility rate) emerges as a marginal factor with much lower order of importance.[5] Therefore, much of the environmental problems seem to be directly linked to environment related factors, some with institutional (role of government, international communities and private sectors), and marginally due to population stress.

As can be seen from a comparable scenario among the neighbouring countries (Table 1), Indian status of environmental sustainability is better than China, but worse than Malaysia, Bhutan, Indonesia, or Sri Lanka. Globally speaking, the highest ESI rank is observed in Finland with a score of 75.1 and lowest being North Korea with 146th rank and a score of 29.2. India ranks hundred and one in the global ordering. Furthermore, among 21 countries with high population density, India ranks seventh, Japan as number one (with ESI score of 57.3) and North Korea as twenty first (with ESI score of 29.2).

A comparison between the ESI and Human Development Index (HDI) ranks indicates that among the neighbouring South Asian countries, as compared to the status of HDIs, the variation in environmental status is much less. Secondly, countries like Bhutan and

Table 1

Environmental Sustainability Index and Human Development Index for India and other Asian Countries

Country	*Environmental Sustainability Index (ESI)*	*ESI ranking on a Global Scale*	*HDI (2004)*	*HDI Ranks*
Malaysia	54.0	23	0.793	59
Bhutan	53.5	43	0.536	134
Indonesia	48.8	53	0.692	111
Sri Lanka	48.5	79	0.740	96
Nepal	47.7	85	0.504	140
Philippines	42.3	97	0.753	83
India	45.2	101	0.595	127
Bangladesh	44.1	114	0.509	138
Pakistan	39.9	131	0.497	142
China	38.6	133	0.745	94

Source: Yale Centre for Environmental Law and Policy (2005) and www.undp.org.

Indonesia with very high ESI seem to be low in their HDI rankings. Likewise, highly populated countries like China seem to be much better in HDI, but very low in ESI ratings. The low level of correlation coefficient (i.e., 0.102) between ESI score and HDI score for these countries suggests that improvement in human development alone is not sufficient to ensure environmental sustainability and vice versa.

Observationally, there are a number of studies to show that wherever peoples' dependency is very high on environmental resources the welfare losses due to environmental degradation is also quite high (Jodha, 1986; Kadekodi, 2004). Essentially, the views go on the lines of complementarity between environmental and natural resources with development. But when it comes to the population related stress factors, one is not sure that the same can be said with equal force.

Some data from selected states for a select set of relevant variables are presented in Table 2 followed by their intercorrelations in Table 3. The correlations however indicate just the direct relevance with no reference to causality, or even indirect causality.

The following observations can be made nonetheless with two qualifications. First, it should be understood that state is not a

Table 2

Selected Population and Environment Related Variables (2001)

	HDI	NSDP per capita (1993-94 prices)	Slum population (%)	Population Below Poverty Line (%	Density (Persons per sq km)	Urbanization (%)	Change in forest cover (%) (1981-2000)	% of habitations affected by various pollutants	Per capita solid waste generated kg/day	Per sq km water pollution load	Per sq km bio Medical waste generation (kg/day)	Degraded land per sq km (2001)	% of wastelands in 2000
Andhra Pradesh	0.416	9982	24.1	15.8	275	27.1	-0.30	20.5	0.364	0.037	0.24	0.44	18.81
Assam	0.386	6157	18.0	36.1	340	12.7	-0.03	58.0	0.223	0.100	0.40	0.38	25.52
Bihar	0.367	3345	23.7	42.6	609	13.4	-0.10	14.5	0.280	0.135	0.42	0.38	12.08
Gujarat	0.479	12975	18.2	14.1	258	37.4	-0.23	4.7	0.451	0.188	0.75	0.64	21.95
Haryana	0.509	14331	16.9	8.7	477	29.0	0.09	1.9	0.276	1.328	0.40	0.94	8.45
Karnataka	0.478	11910	9.3	20.0	275	34.0	-0.02	5.0	0.376	0.409	4.94	0.59	10.87
Kerala	0.638	10627	15.9	12.7	819	26.0	0.38	7.2	0.393	2.171	4.97	0.50	3.73
Madhya Pradesh	0.394	7003	13.7	37.4	175	25.0	-0.14	3.6	0.316	0.217	0.10	0.47	15.72
Maharashtra	0.523	15172	25.8	25.0	314	42.4	-0.26	0.6	0.378	0.335	0.64	0.64	17.38
Orissa	0.404	5187	19.9	47.2	236	15.0	-0.19	56.2	0.366	0.838	0.23	0.50	13.71
Punjab	0.537	15390	23.6	6.2	482	34.0	0.08	13.4	0.312	2.612	0.73	0.64	4.42
Rajasthan	0.424	7937	23.8	15.3	165	23.4	-0.46	33.0	0.355	0.597	0.15	1.09	30.87
Tamil Nadu	0.531	12779	18.7	21.1	478	43.9	-0.01	20.2	0.467	1.802	0.94	0.29	17.7
Uttar Pradesh	0.388	5770	21.1	31.2	424	21.0	-0.33	1.6	0.381	0.826	0.40	0.45	13.17
West Bengal	0.472	9778	27.8	27.0	904	28.0	-0.09	1.8	0.321	3.622	1.54	0.48	6.44

Table 3

Correlation Matrix

	HDI	*NSDP*	*SLUM*	*BPL*	*DEN-SITY*	*URBAN*	*FCHANGE*	*HABITA-TION*	*SWASTE*	*WPLOAD*	*MWASTE*	*DLAND*	*WLAND*
HDI	1	0.762	-0.122	-0.689	0.456	0.646	0.668	-0.394	0.401	0.565	0.595	0.154	-0.484
NSDP		1	-0.029	-0.801	0.063	0.876	0.294	-0.470	0.334	0.333	0.233	0.335	-0.242
SLUM			1	0.026	0.274	-0.035	-0.408	0.042	-0.099	0.283	-0.516	0.051	0.078
BPL				1	-0.120	-0.614	-0.265	0.421	-0.267	-0.330	-0.289	-0.519	0.150
DENSITY					1	-0.018	0.613	-0.325	-0.103	0.766	0.391	-0.277	-0.685
URBAN						1	0.082	-0.553	0.636	0.235	0.208	0.127	-0.095
FCHANGE							1	-0.146	-0.141	0.495	0.613	-0.217	-0.692
HABITATI								1	-0.271	-0.241	-0.288	-0.107	0.482
SWASTE									1	0.045	0.256	-0.087	0.076
WPLOAD										1	0.270	-0.019	-0.641
MWASTE											1	-0.085	-0.470
DLAND												1	0.207
WLAND													1

Note: HDI: Human Development Index (2001); NSDP: Net State Domestic Product (Per capita in 1993-94 prices) (2001); SLUM: Population living in slums – 2001(Percentage); BPL: Population Below Poverty Line – 2001 (Percentage); DENSITY: Population Density – 2001 (Persons per sq. km); URBAN: Degree of Urbanization – 2001 (Percentage); FCHANGE: Forest cover change between 1981 and 2001 (Percentage); HABITATION: Habitations affected by various pollutants – 2001 (Percentage); SWASTE: Solid Waste – 2001 (Per capita Kg/day); WPLOAD: Water pollutant loads – 2001 (Per sq. km; MWASTE: Bio-medical waste – 2001(Kg/day/per sq. km); DLAND: Degraded Land – 2001 (Per sq. km); WLAND: Waste land – 2000 (% to geographical area)

representative unit for analysing the linkages and nexus between population and environment. Second, there are a number of additional factors and variables that can influence both the population dynamics and the environmental status. Some examples are degree of governance, law and order, community participation in health environmental matters and so on. Third, some of the inferences made here are based on the direct measures of linkages (by looking at the partial correlation coefficients), though it is well recognized that indirect linkages are also equally important (to be based on an econometric model of the kind shown in Figure 1).

- It can be safely inferred that states with high HDI seem to be blessed with lower slum population and people below poverty, and low level of habitats affected by various pollutants. But they are highly responsible for the generation of high rates of biomedical and solid wastes, and water pollutants. In other words, the state of human development is also responsible for a different type of consumption patterns leading to the generation of such environmentally sensitive matters.
- Two major indicators of population dynamics are selected. They are: population density and degree of urbanization. While it is expected that states with high population pressure may end up with high levels of environmental degradation, it is not seen clearly with the data for the latest year. For instance, while water pollution loads and generation of biomedical wastes are high with population densities, the extent of solid waste generation or habitats affected are not are high (rather found to be low).
- States with high population densities seem to have taken environment degradation bit more seriously, as can be seen from the increasing forest covers, low levels of waste and degraded lands.
- The extent of solid and medical waste generation as well as water pollutants seem to be high with high degree of urbanization. Strangely enough, urbanization also seems to have reduced the extent of habitations affected by pollutants, and reduced the extent of waste lands.

In summary, one does not see much of direct linkages showing adverse environmental situations, substantially attributable to population pressure or degree of urbanization.

Towards a Better Understanding of the Nexus between Population and Environment

Much of human development has taken place in terms of provisioning and enabling better health, education facilities, and income earning opportunities in some of the states in India. They act as 'pressures' on various socio-economic and environmental attributes (or variables). Examples are, with better health provisioning, expectation of life can improve, degree of urbanization may improve, per capita income may increase, demand for groundwater may increase, forest degradation may take place, solid and medical wastes may increase and so on. Then the 'impact' and 'outcomes' on improved, worsened, or changed socio-economic welfare and environmental status are important indicators to examine. Equally important are to examine the response by the state or governments and the feedback of all these acting as pressures once again. Some examples of responses and state actions are, meeting demands on provisioning of better health care, improved educational facility, income earning opportunities, monitoring pollution and hazardous environmental emissions and effluents, regulating ground and other water resources, introduction of forest protection committees under joint forest management, encouraging water user association and so on. Such a cyclical nexus between human development and environment is shown in Figure 2.

Figure 2

The PASIR Framework*

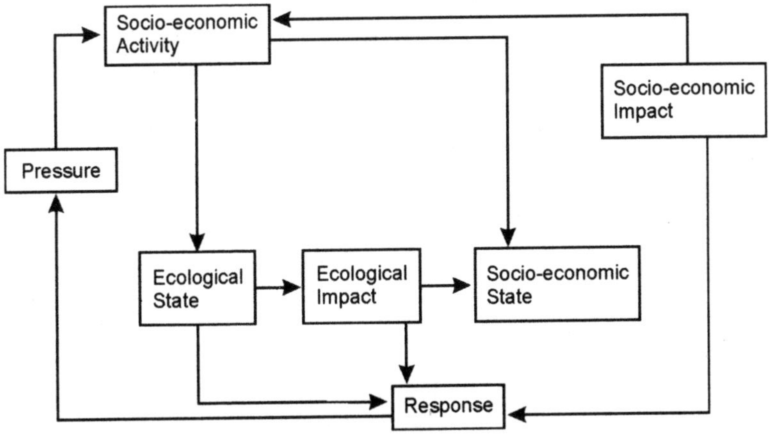

*PASIR: Pressure Activity State Impact Response

Most of the pressure variables are directly working on the human development front, more as provisioning and enabling the facilities for better health, activity status and income prospects. Some indicative factors and variables so grouped are shown below.

Indicative pressure indicators are:

Developmental	*Health Population*	*Social*	*Environmental*
Poverty ratio	Population density	Literacy rates	Area under Forest
Per capita income	Degree of urbanization	PHCs* and hospitals	Amount of CPRs* available
	Expectation of life at birth	Immunization Family welfare	Access to water and fuels
	Infant mortality	Counseling centres	Wastelands
	Child mortality	Overall HDI	Slums
			Land degradation
			Solid waste generation
			Pollutants and their loads

* PHC: Primary Healthcare Centres
CPRs: Common Property Resources

Examples of directly measured relevant impact and outcome variables are:

- Status of groundwater resources
- Changes in forest cover
- Water pollution loads
- Fuelwood collections from CPRs
- Production of biomedical wastes
- Hazardous and solid waste generation
- Number of slums
- Percentage of people and households living in slums
- Degree of urbanization
- Undernourishment
- Incidence of diseases
- Mortality rates
- Physical Quality of Life Index and gender ratios

Against these, the response from the government (i.e., state) can be tracked using some of the following variables:

- Number of defaulters to comply with environmental regulations
- Disposal of solid waste by different agencies
- Creation of forest protection committees
- Creation of water user associations
- Monetary and fiscal incentives
- Fiscal instruments: taxes, subsidies, user charges, etc.
- Promotion of NGOs and User groups

Such a model and the analysis of its dynamic process also make high demand on data. Therefore, it may be feasible and logically necessary to carry out the study on smaller geographical entities such as a watershed or at best at a basin level. Some studies on these lines were conducted for Palamau district of Jharkhand (Chopra and Kadekodi, 1999). More studies based on different agro-climatic regions are hoped.

Annexure 1

Relative Importance of Different Components of ESI for India

Air Quality	–0.98
Biodiversity	–0.62
Land	–0.36
Water Quality	–0.96
Water Quantity	–0.75
Reducing Air Pollution	–0.28
Reducing Ecosystem Stress	0.32
Reducing Population Stress	0.12
Reducing Waste and Consumption Pressures	0.35
Reducing Water Stress	–0.27
Natural Resource Management	–0.25
Environmental Health	0.08
Basic Human Sustenance	–0.04
Reducing Environment Related Natural Disaster Vulnerability	–0.37
Environmental governance	0.04
Eco-efficiency	0.00
Private Sector Responsiveness	0.50
Science and Technology	–0.28
International Collaborative Efforts	0.67
Greenhouse Gas Emissions	–0.37
Reducing Transboundary Environmental Pressures	0.92

Source: Yale Centre for Environmental Law and Policy (2005).

Note: These scores represent the average of the standardized z-scores of the variables that comprise the indicators. Higher numbers represent higher levels of performance; scores near the central axis are closer to the mean score for that indicator; and negative scores represent below average performance for the complete set of 146 countries included in the ESI.

Notes

1. Duraiappah et al. (2000) formulate such a dynamic model, which in brief termed as PASIR (Pressure-Activity-State-Impact-Response) model.
2. At this stage one may accept the definition of sustainable development as given by the World Commission on Environment and Development (1987).
3. www.eia.doe.gov/emeu/cabs/indiaenv.htm

4. See Annexure 1 for the grouped variables, and their relative performance and importance towards ESI.
5. Annexure 1 shows the relative importance of the factors influencing the ESI.

References

Agarwal, Anil and Sunita Narain (1997): "Dying Wisdom: Rise, Fall and Potential of India's Traditional Water Harvesting System", Centre for Science and Environment, New Delhi.

Anonymous (2005): "Securing India's Future: Final Technical Report of the National Biodiversity Strategy Action Plan", *Kalpavriksha*, Pune.

Arrow, K.J., B. Bologna, R. Costanza, P. Dasgupta, C. Folke, C.S. Holling, B.O. Jansson, S. Levin, K.G. Maler, C. Perrings and D. Pimentel (1995): "Economic Growth, Carrying Capacity, and Environment", *Ecological Economics*, 15(2), pp. 91-5.

Beckerman, W. (1992): "Economic Growth and the Environment: Whose Growth? Whose Environment?", *World Development*, 20(4), pp. 481-96.

Bojö,J., J. Bucknall, K. Hamilton, N. Kishor, C. Kraus and P. Pillai (2001): "Sourcebook on Poverty, Environment, and Natural Resources for the Poverty Reduction", Strategy Papers (draft).

Boulding, Kenneth E. (1966): "The Economics of the Coming Spaceship Earth", In H. Jarrett (ed.), *Environmental Quality in a Growing Economy: Resources for the Future,* Johns Hopkins University Press, Baltimore, MD, pp. 3-14.

Chopra, K. and Gopal K. Kadekodi (1999): *Operationalising Sustainable Development: Economic-Ecological Modelling for Developing Countries*, Sage Publications, New Delhi.

Dasgupta, P. (2003): "Population, Poverty, and the Natural Environment" in, K.G. Maler and J. R. Vincent (eds.). *Handbook of Environmental Economics,* Vol. I, Edward Elgar, Cheltenham, pp. 191-247.

Dasgupta, P. and K.G. Maler (1994): "Poverty, Institutions and the Environmental-Resource Base", *World Bank Environment Paper No. 9*, The World Bank, Washington, D.C.

Datt, Divya, S.C. Garg, Gopal K. Kadekodi, K.K. Narang, Dharmendra Sharma and J.P. Singh (2004): "Environmental Fiscal Reforms in India: Issues and Some Steps Forward", in *Environmental Fiscal Reform for Sustainable Development and Poverty Reduction*, OECD Publication.

Duraiappah, A.K. (1998): "Poverty and Environmental Degradation: A Review and Analysis of the Nexus". *World Development*, 26(12), pp. 2169-79.

Duraiappah, A.K., G. Ikiara, M. Manundu, W. Nyangena and R. Sinange (2000): "Land Tenure, Land Use, Environmental Degradation and Conflict Resolution: A PASIR Analysis for the Narok District, Kenya", *CREED Working Paper No. 33,* IIED, London, 2000.

Grossman, G.M. and A. Krueger (1995): "Economic Growth and Environment", *Quarterly Journal of Economics*, 112, pp. 353-78.

– (1996): "The Inverted U Curve: What Does It Mean?" *Environment and Development Economics*, 1, pp. 119-22.

Jodha, N.S. (1986): "Common Property Resources and Rural Poor in Dry Regions of India", *Economic and Political Weekly*, Vol. 21, No. 27, pp. 169-181.

– (2000): "Wastelands Management in India: Myths, Motives and Mechanisms", *Economic and Political Weekly*, February 5, pp. 466-473.

Kadekodi, Gopal K. and L. Venkatachalam (2005): "Human Development, Environment and Poverty Nexus in India", report submitted to UNDP, New Delhi.

Kadekodi, G.K. (2004): *Common Property Resource Management: Reflections on Theory and Indian Experience*, Oxford University Press, New Delhi.

– (2001): "Environmental Degradation and Resilience to Human Development: A Case Study of Karnataka", In A.P. Bali (ed.), *Refashioning the New Economic Order: Karnataka in Transition,* ICSSR and Rawat Publications, New Delhi.

Lopez, R.E. (1997): "Where Development can or cannot go: The Role of Poverty-Environment Linkages", *Annual World Bank Conference on Development Economics*, The World Bank, Washington D.C. pp. 285-306.

Markandya, A (1998): "Poverty, Income Distribution and Policy Making", *Environmental and Resource Economics,* Vol. 11 (3-4), pp. 459-72.

– (2001): "Poverty, Environment and Development", in, H. Folmer, H.L. Gable, S. Gerking and A. Rose (eds.), *Frontiers of Environmental Economics*, Edward Elgar, Cheltenham

Nadkarni, M.V. (2001): "Poverty, Environment and Development in India", in A. Heyes and M.V. Nadkarni (eds.), *Poverty, Environment and Development,* UNESCO, Bangkok.

– (2000): "Poverty, Environment, Development: A Many Patterned Nexus", *Economic and Political Weekly*, Vol. 35 (14), pp. 1184-90.

Seldon, M. and D. Song (1994): "Environmental Quality and Development: Is there a Kuznets Curve for Air Pollution Emission?", *Journal of Environment and Management*, Vol. 27, pp. 147-62.

Shafik, N. and S. Bandyopadhyay (1992): "Economic Growth and Environmental Quality: Time Series and Cross-country Evidence", *World Bank Background Report.*

Yale Centre for Environmental Law and Policy (2005): "2005 Environmental Sustainability Index: Benchmarking National Environmental Stewardship", In: www.yale.edu/esi

World Commission on Environment and Development (WCED) (1987): *Our Common Future*, Oxford University Press, New Delhi.

2 Understanding the Linkages between Population Growth, Depletion of Natural Resources and its Implications for Women: A Focus on the EAG States

Kamla Gupta and Mohua Guha

Human activities modify the natural environment, and the number and distribution of people affect the scale and nature of their activities. Meeting the challenge of population size, change and imbalances in distribution, in the context of continued degradation of critical resources – forests, soils, air, water, flora and fauna – is, therefore, one of the most important tasks facing humankind. The relation between population, economic activity and environment is not simple as it is constantly changing, responding to changes in population growth, production, environment and government policies. Most of the environmental changes during the twentieth century have been a product of the efforts of humans to secure improved standards of food, clothing, shelter, comfort and recreation.

The central role of the absolute size of a population in generating environmental stress was first posited by Ehrlich in 1968 in the form of an equation relating environmental impact to the product of population size, affluence and technology. The depletion of both renewable and non-renewable resources may be caused by rapid changes in population size and abrupt changes in activity or by inadequate ecological adaptation. There are manifold examples of higher densities of population transforming landscape which inevitably have profound deleterious effects upon the environment. Increasing population pressure in marginal areas, especially semi-arid lands, mountains and some humid tropical lands have led to greater demands for food, fuel wood and water, changes in land use, agricultural systems and production and has accelerated environmental degradation of these fragile ecosystems. Population pressures are not easily quantified as they are greatly affected by ecological fragility of areas, land use systems, awareness of the need to conserve certain resources and their environmental importance. Evidences of environmental degradation due to inappropriate agricultural practices, deforestation, desertification and water shortage are widespread phenomena which tax the ingenuity of the rural poor. The poor, in general, also do not have the means of undertaking resource conservation measures, particularly soil conservation measures on which depends their long-term development (Tabah, 1992). Thus, the study of population-environment relationships should concern itself with interactions in both directions: with the feedback processes which may evolve to produce spiraling crises for the population itself; and with the identification of interventions which might prevent destructive interactions.

It is widely accepted that development processes over the last several years have neglected both women and environmental issues. Along with the growing perception and articulation of this neglect is the perception of the link between the two, that is, of gender perspectives within environmental concerns. That woman plays a predominant role in activities relating to natural resources and that the degradation of these resources in turn has a disproportionate effect on women is fairly well established by now. There is also a strong line of thought that endorses women's special consciousness of ecological issues which supposedly makes them better managers of natural resources. Depletion of resources, in turn, adversely affects the availability of products such as fuel and fodder, the time required to obtain these products and therefore women's workload, household nutrition and food intake. Hence, the conservation of biodiversity of ecological system can strengthen women's autonomy and status by enabling them to continue their productive and reproductive functions effectively.

The Empowered Action Group (EAG) formed during 2001 focused on specific states, viz., U.P., M.P., Rajasthan, Bihar, Orissa, Uttaranchal, Jharkhand and Chhattisgarh to implement area-specific programmes. Bihar, U.P., M.P., Rajasthan and Orissa have been among the most populous states of India throughout the twentieth century. Uttaranchal, Jharkhand and Chhattisgarh being part of former U.P., Bihar and M.P. respectively display the same demographic and socio-economic traits as their predecessors. The other important features which distinguish these states from many other states of the country are their fragile ecosystems like mountainous periphery of the Garhwal Himalayas, difficult terrain, inaccessible rain forest, semi-arid areas, inadequate infrastructure (like roads and rails) in addition to high density of population. Categorically, it can be stated that the benefits of planned development in these states have been offset by rapid population growth and only a marginal part of the development contributes towards improving the existing indices of development. This is evident from the shrinkage of forest cover, low per capita availability of food grains and socio-economic backwardness. Statewise variation in terms of development has created regional heterogeneity. Keeping these above facts in view, an attempt is made in the paper to understand the interrelationship between population growth, use and exploitation of natural resources and its implications for the women folk in the context of the EAG states in India.

Data and Methodology

The data for the present study has been gleaned from multiple sources in view of the array of different parameters involved to assess rate of population growth, depletion of natural resources and understanding the implications on women's life. The different sources are Census of India, 2001; Population of India, 2001; Agricultural Statistics at a Glance, August 2004; NFHS-2 (1998-99); State of Forest Report, 2001; Compendium of Environment Statistics, 2003; Profiles of Districts, CMIE, October 2000, Monthly Review, CMIE, 2005 for the states of M.P., U.P., Rajasthan and Uttaranchal. Proportions and rates have been calculated to bring out the relationship between all the selected indicators.

Results and Discussion

Dynamics of Population

The large population base and fast growing population has resulted into heavy pressure on the land and other natural resources in all the EAG states. As a whole, the EAG states currently possesses 45 per cent of the country's total population. U.P. has a population of 166 million, followed by Bihar (82.9 million), M.P. (60.3 million) and Rajasthan (56.5 million). As regards to the natural growth rate of population during the period of 1951-2001, Bihar, Jharkhand, M.P., Rajasthan and

Table 1

Percentage Decadal Growth Rate of Population for the EAG States, 1951-2001

States	*1951-61*	*1961-71*	*1971-81*	*1981-91*	*1991-2001*
Bihar	19.8	20.9	24.2	23.4	28.4
Chhattisgarh	22.8	27.1	20.4	25.7	18.1
Jharkhand	19.7	22.6	23.8	24.0	23.2
M.P.	24.7	29.3	27.2	27.2	24.3
Orissa	19.8	25.1	20.2	20.1	15.9
Rajasthan	26.2	27.8	33.0	28.4	28.3
U.P.	16.4	19.5	25.4	25.6	25.8
Uttaranchal	22.6	24.4	27.5	24.2	19.2
India	21.6	24.8	24.7	23.9	21.3

Source: Census of India, 2001.

U.P. have registered higher growth rate than the rate for country. Table 1 lends a microscopic view into the trend of population growth over the period 1951 to 2001. Except Bihar and U.P., the growth rate for the rest of the states shows a declining trend for the period 1991-2001. U.P., with a three-fold increase in its population from 48.6 million in 1901 to 166.1 million in 2001 remains the most populous state in India. Bihar records the highest growth rate (28.4 in 2001), which has increased by five per cent points from the last decade. Chhattisgarh depicts a substantial decrease in the decadal growth rate in 2001. The growth rate has been consistently declining in Orissa over the past few decades and is the lowest (15.9 in 2001) among the EAG states.

Table 2 presents the density of population (persons per sq km) for the years 1991 and 2001. The population density is as high as 880 in Bihar, followed by 689 in U.P., 196 in M.P. and 165 in Rajasthan. All the states have recorded an increase in density per sq km which signifies that additional population have to draw the means of sustenance as well as to meet the growing demands from the land resources, which is finite. In the past fifty years, the population growth in U.P. has resulted into an increase in density from 215 in 1951 to 689 persons per sq km in 2001. The addition of persons per sq km in Bihar during the period 1991-2001 (195 persons) is the highest among the Indian states followed by U.P. (addition of 141 persons per sq km in 1991-2001). The increase in density in Bihar and U.P. has been about thrice as

Table 2

Population Density in the EAG States, 1991-2001

(persons per sq km)

States	*2001*	*1991*	*Increase in density (per sq km) in 2001*
Bihar	880	685	195
Chhattisgarh	154	130	24
Jharkhand	338	274	64
M.P.	196	158	38
Orissa	236	203	33
Rajasthan	165	129	36
U.P.	689	548	141
Uttaranchal	159	133	26
India	324	256	68

Source: Census of India, 2001.

much higher as the national average (addition of 57 persons per sq km). There are also wide district level variations in the density of population in all these states (see Appendices).

The immediate consequence of such rapid population growth has been the decline in the size of the operational landholding[1] in the concerned states. The average size of landholding (Table 3) had reduced from around 2.28 hectares in 1970-71 to 1.41 hectares in 1995-96 in India – a decline of 38 per cent during this period. Similar is the situation in all the EAG states. Rajasthan and U.P. has witnessed almost one fourth reduction in the average size of operational holding over the period from 1970-71 to 1995-96. In M.P. it has declined by 43 per cent (from 4.00 hectares to 2.28 hectares) and in case of Bihar also there has been a substantial decline (50%) over the period. This is largely due to the fact that as a result of increase in population, the number of operational holdings has decreased owing to progressive reduction of agricultural land due to urbanization and the government policy of industrialization of backward areas. Again, data about the percentage distribution of the number and operated area of holdings indicate wide variations. On the one extreme, the states like Bihar and U.P. have a very high proportion of marginal holdings (ranging from 65 to 80 per cent of the total) and accounting for a very large proportion of area (ranging from 21 to 40%). On the other states like

M.P., Rajasthan and Orissa have proportion of marginal holdings in the range of 24 to 33 per cent which accounts for only 3 to 6 per cent of the operated area. Thus, rapid population growth has made majority of the state landholdings unviable, having crossed the barrier of subsistence holdings.

Table 3

Average Size of Operational Holding in the EAG States

(average size in hectares)

States	*1970-71*	*1990-91*	*1995-96*
Bihar	1.50	0.93	0.75
M.P.	4.00	2.63	2.28
Orissa	1.89	1.34	1.30
Rajasthan	5.46	4.11	3.96
U.P.	1.16	0.90	0.86
India	2.28	1.57	1.41

Sources: 1. Ministry of Agriculture, *Agricultural Situation in India*, October, 1981. *Agricultural Statistics at a Glance ,1995.*
2. *Agricultural Statistics at a Glance*, August 2004.

The acute pressure of population on land resources in case of the EAG states is of particular significance because of the fact that even in 2001 nearly three-fourths of the states' main workers were engaged in primary activities (Registrar General of India, 2001). In most of these states, agriculture is the main stay of the population. Currently, the agricultural share accounts for 34 and 32 per cent respectively of the State Gross Domestic Product (at current prices) during 2000-2001 for Bihar and U.P. For the other states, agriculture adds nearly one-fourth of state's total income. The industrial backwardness of these states is reflected in the fact that Rajasthan, M.P. and U.P. have only 4.1, 2.4 and 7.1 per cent respectively of India's total registered factories which accounts for as low as 3.2, 3.97 and 6.97 per cent respectively of the total value of output by manufacturing industries or the secondary sector (Monthly Review, Centre for Monitoring Indian Economy (CMIE), September 2005 for the states of M.P., U.P. and Rajasthan). The level of urbanization in these eight states is also significantly lower when compared to the country's average level of urbanization (27.81% in 2001). The per cent of urban to total population is around 26, 23, 21, 15 and 10 respectively in M.P., Rajasthan, U.P., Bihar and Orissa (see Appendices). In the newly created states of Chhattisgarh, Jharkhand

and Uttaranchal also, the level of urbanization is as low as 20, 22 and 26 per cent respectively. Out of the total 70 districts (Census of India, 2001) of U.P., 45 districts depict less than 20 per cent of urban population and as many as 21 districts have less than 10 per cent level of urbanization. In Bihar, the level of urbanization is less than 10 per cent in 27 districts (out of a total of 37 districts). Similarly in 34 districts of M.P., the level of urbanization is below the state average. Therefore, the rising pressure of population in all these EAG states is impinging on the already finite land resources which, in turn, are acting as an impediment for industrial and urban development.

Status of Food Production and Resource Depletion

The population growth of these respective states has also nullified the progress attained by the country in food production. The production of food grains has decreased from 44.14 million tonnes in 2001-02 to 36.30 million tonnes in 2002-03 in the state of U.P. (Table 4). In Bihar, there has been a slight increase in percentage share of the total production of food grains (from 5.49 in 2001-02 to 5.90 in 2002-03). In all the other states, both production and yield is showing a declining trend over the same period. This lower per capita availability of food is reflected in the widespread incidence of undernutrition and anaemia in all the eight EAG states. According to the National Family Health Survey (NFHS-2) 1998-99, both acute and chronic undernutrition are found to be very high in these states. Levels of anaemia are substantially high in all the states, and particularly in Bihar and Orissa almost two-thirds of the women suffer from iron deficiency. In most of these states the prevalence of anaemia is as high as 70 per cent among children (aged 6-35 months). Table 5 presents a detailed overview of the nutritional status of children under age three years for the five major EAG states (Jharkhand, Chhattisgarh and Uttaranchal being part of former Bihar, M.P. and U.P. respectively). Again the incidence of undernutrition is higher in rural areas compared to their urban counterparts. There exist wide intra-state variations in availability of per capita food grains in most of these populous states, thereby, indicating acute food shortage. Thus, it is necessary to increase food production, not only to fulfil the subsistence need of the growing population but also to provide supplementary food in order to achieve the desired nutritional status of the population of the EAG states.

Table 4

Area, Production and Yield of Food Grains in the EAG States, 2002-03

States	*Area (in million hectares)*	*% of Total Area*	*Production (in million tonnes)*	*% of Total Production*	*Yield (in kg per hectare)*
U.P.	17.90	16.05	36.30	20.84	2028
Bihar	6.88	6.17	10.27	5.90	1493
M.P.	10.25	9.19	9.75	5.60	951
Rajasthan	8.66	7.77	7.57	4.35	874
Orissa	5.04	4.52	3.56	2.04	705
Chhattisgarh	4.90	4.39	3.16	1.81	645
Jharkhand	1.89	1.70	2.68	1.54	1421
Uttaranchal	1.03	0.92	1.56	0.90	1511
India	111.50	-	174.19	-	1562

Source: Agricultural Statistics at a Glance, August 2004.

Table 5

Nutritional Status of Children in the EAG States

States	*Weight-for-age*[2]		*Height-for-age*[3]		*Weight-for-height*[4]	
	% below -3SD	*% below -2SD**	*% below -3SD*	*% below -2SD**	*% below -3SD*	*% below -2SD**
Rajasthan	20.8	50.6	29.0	52.0	1.9	11.7
M.P.	24.3	55.1	28.3	51.0	4.3	19.8
U.P.	21.9	51.7	31.0	55.5	2.1	11.1
Bihar	25.5	54.4	33.6	53.7	5.5	21.0
Orissa	20.7	54.4	17.6	44.0	3.9	24.3
India	18.0	47.0	23.0	45.5	2.8	15.5

Note: Each index is expressed in standard deviation units (SD) from the median of the International Reference Population.

* Includes children who are below -3 SD from the International Reference Population.

Source: NFHS-2, India, 1998-99.

This growing demand of food is difficult to be fulfilled without seriously damaging the environment. The soaring densities of population over the years are eventually creating pressure on the existing land resources and there is hardly any scope to bring additional land under cultivation. Whatever expansion has been achieved in food production has been brought through clearing of vegetative cover along with increased use of chemical fertilizers, irrigation, pesticides, etc. The over utilization of such agricultural inputs have been found to be environmentally unsustainable. Almost in all of these states deforestation and degradation of top soils have given rise to arid conditions, thereby reducing the food producing potential of the soil. Present estimates show that the forest cover of the country is less than the goal laid down in the policy (one-third of the geographic area) and trend over the years depict that the EAG states, in particular, are losing forest cover more rapidly compared to the country, in general. It is evident from Table 6 that hardly 6, 9 and 7 per cent respectively of the geographical area of the states of Bihar, Rajasthan and U.P. are under forest cover whereas a minimum of 60 per cent is required to maintain the ecological balance and environmental stability. Except Chhattisgarh, M.P., Orissa and Uttaranchal, all the other states have forest cover even below the current national average (23%), which is inadequate. Dense forest constitutes about 78 per cent of the total area under forest cover in Uttaranchal followed by Chhattisgarh, U.P. and Bihar. Open accessible forests in the proximity of the settlement areas are reported to be degrading faster due to collection of fuel wood, cattle grazing, etc., as compared to forests located in remote localities with sparse population growth. Again the distribution of forest areas within each of these states is also very uneven (see Appendices). In reality, not more than 35 per cent of the surface area of the Himalayan districts of U.P. and Uttaranchal is now covered with dense forests (Gupta and Joshi, 1987). However, a study conducted in 1998 by the Forest Survey of India in collaboration with the Indian Bureau of Mines, Nagpur (GOI, 2003) have concluded that out of 90,695 hectares of leased area in the three mineral rich states of Bihar, M.P. and Orissa, 53,217 hectares is under forest cover of which 71 per cent is under dense forest and 29 per cent is under open forest. Yet the latest satellite data confirms that India is losing 1.3 million tonnes of forests per year, nearly eight times the annual rate put out by the Forest Department (CSE, 1985). Therefore, it is necessary to create massive people's movement through involvement of village communities living close to the forest in protection and development of forests, especially through female intervention in a more comprehensive way.

Table 6

Area under Forest by Type of Forests in the EAG States, 1999

States	*Percentage*			
	Forest to total area	*Dense forest*	*Open forest*	*Scrub*
Bihar	6.45	57.72	40.19	2.09
Chhattisgarh	43.85	66.87	32.78	0.35
Jharkhand	29.61	49.92	45.95	4.13
M.P.	30.89	54.99	40.74	4.28
Orissa	37.34	51.21	38.20	10.59
Rajasthan	9.49	29.69	47.18	23.13
U.P.	6.98	62.15	33.15	4.70
Uttaranchal	64.81	77.53	20.03	2.44
India	23.38	57.66	35.79	6.55

Source: State of Forest Report, 2001.

Studies show that deterioration of the environment affects the life and work of the people in all areas and among all socio-economic groups. However, the situation is particularly critical in case of marginal ecosystems like mountains, arid and semi-arid areas and coastal plains. The people inhabiting such areas are the first to face the total social and economic disruption because of environmental deterioration. Again, it is the poor in particular who bear the brunt of this environmental damage more as they do not possess either land or cash resources to counteract the environmental problems. Most of the poor households in the rural areas of these states depend on their immediate environment for food supplies, fuel, fodder, etc. The community forests, pastures, ponds, rivers, lakes, etc., usually acts as village commons, which are important sources of employment and income for the rural households. However environmental degradation and developmental activities have resulted in the shrinkage of these common property resources (CPRs) too, both in terms of physical extent and quality (Jodha, 1985). This leads to severe rural poverty and out-migration, by pushing people, who are dependent on CPRs, to other areas.

Environmental Degradation and Its Implications for Women

The impact of environmental degradation has been found particularly deleterious on women. Average women experience deterioration in their living conditions because of either depletion or commercialized intervention in exploitation of natural resources, which limits her interaction with nature (Acharya, 1997). Since women are subordinate to men virtually in all the societies, they are subjugated to a variety of social and economic discrimination right from birth. In a traditional Indian family, a girl child is discriminated against a male child even in provision of such basic necessities such as food, education and Medicare. The discrimination in nutrition and other health care utilization is reflected in the gender differentials in the infant and child mortality in the context of India. The sex ratio of infant and child mortality is adverse to girls, in spite of the fact that they are biologically stronger than boys (Basu, 1991; Jejeebhoy, 1990). The negative, synergistic effect of femaleness and poverty on nutritional status is further borne by the fact that nutritional equity between males and females is lower in the states of U.P., Bihar and Rajasthan which eventually improves towards the south (World Bank, 1991). A study on energy consumption by the two sexes showed that average men consume nearly 800 calories more than what they spend, while women take 100 calories less (Batliwala, 1983). Over 75 per cent of pregnant and 80 per cent of lactating women in sedentary activity and 100 per cent of moderately active pregnant or lactating women in India fail to consume recommended levels (World Bank, 1991). Continued impoverishment of the poor households owing to environmental susceptibility signifies a further cut in the already meagre food intake by the females in India.

Deterioration of the physical environment, leads to increasing work burden on women. In the sexual division of labour in the Indian society, women are burdened by double work load (at home and outside) and have to undertake all those 'domestic chores' which are labour intensive, time consuming, without any cash remuneration and monotonous. Besides bearing and rearing of children, women are also entrusted with the tasks of serving and catering to the needs of the male family members, the aged and also the children. It is observed that more than half of women's time is spent on such basic activities as cooking food, fetching water, collecting fuel and fodder and taking food to farms. On an average a woman is found to spend about 7.5 to 10 hours on all such activities (Batliwala, 1983). Depletion of easily available natural resources further aggravates the situation by increasing the time and cash costs of securing fuel wood necessary to transform raw foodstuffs into meals. Numerous recent studies document the increasing amount of time and effort women must spend in this task and its impact on nutrition, health and the schooling of children (Agarwal, 1985a; Agarwal, 1981; Batliwala 1983; CSE 1985; Nagbrahman and Sambrani 1983). NSS data also shows that the amount of time women spend in 'free collection' of fuel, fodder and other materials is inversely related to the size of landholding. Their plight is further worsened if the male members out-migrate, which generally happens when environment deteriorates. In such conditions, women have to take up most of those tasks also which are generally performed by males. Because of heavy out-migration in the villages of Garhwal in U.P., women are forced to devote more time and efforts into agriculture (Swaminathan, 1985).

Conclusion

In all the EAG states of India, the rapid population growth has put severe pressure on land and forest resources, thereby resulting into scarcity of natural resources. Many areas of these states have been severely affected by environmental problems like deforestation, degradation and salinization of productive soil due to inappropriate agricultural practices generating acute food shortage, sometimes culminating into famine-like conditions. Clearly, deforestation has affected the domestic economy of poor households with greater severity, and its impact has fallen primarily on the womenfolk since they shoulder the main responsibility for fuel and fodder collection for the household. The scarcity of everyday resources within the periphery of the community dwellings demands unending work burden with added time allocation for carrying out such activities. This adversely affects women's health and welfare, which has serious implications on many other aspects of women's development, like utilization of educational, health and family welfare services (Gupta, 1997). Though there are numerous social and cultural factors that inhibit women to utilize these services, but extraordinary work burden in the face of deteriorating environment hardly leaves any time and energy for them to seek health care and participate in other programmes. Women's role in environmental protection is also highly significant since they are the main beneficiaries of the natural resources. It is, therefore, imperative to frame policies and programmes keeping women's interests and well-being at the centre stage and the best way of ensuring protection is to take care of the environmental quality and maintain ecological stability.

Appendix I

District Profile of Orissa for the Year 2001

	Density	*Decadal Growth Rate (1991-2001)*	*Female Work Participation Rate*	*% Urban to Total Population*
Orissa	236	15.94	24.66	14.99
Anugul*	179	18.55	26.66	13.90
Balangir	203	8.52	28.40	11.54
Baleshwar	532	19.24	11.44	10.89
Bargarh*	231	11.47	31.26	7.69
Baudh*	121	17.45	35.54	4.83
Bhadrak*	532	20.47	8.07	10.58
Cuttack	595	14.00	13.76	27.39
Debagarh*	93	17.02	38.58	7.33
Dhenkanal	240	12.46	15.11	8.71
Gajapati*	120	14.02	49.72	10.19
Ganjam	385	16.01	30.92	17.60
Jagatsinghapur*	634	13.15	11.49	9.88
Jajapur	560	17.08	6.84	4.49
Jharsuguda*	245	15.13	22.64	36.47
Kalahandi	169	17.99	35.86	7.50
Kandhamal	81	18.60	42.20	6.80
Kendrapara*	492	13.25	9.99	5.69
Kendujhar	188	16.79	28.05	13.64
Khordha*	667	24.79	8.82	42.92
Koraput	134	14.41	40.42	16.81
Malkangiri*	87	13.71	42.29	6.87
Mayurbhanj	213	17.89	39.89	7.00
Nabarangapur*	194	20.26	42.43	5.78
Nayagarh*	222	10.39	11.01	4.29
Nuapada*	138	13.00	36.79	5.66
Puri	432	14.80	7.67	13.58
Rayagada*	118	15.27	16.61	13.89
Sambalpur	141	14.17	35.65	27.12
Sonapur*	232	13.39	32.75	7.39
Sundargarh	188	16.26	28.88	34.37

* *Newly created districts in the Census of India, 2001.*

Appendix II

District Profile of Chhattisgarh for the Year 2001

	Density	*Decadal Growth Rate (1991-2001)*	*Female Work Participation Rate*	*% Urban to Total Population*
Chhattisgarh	154	18.06	40.04	20.09
Bastar	87	18.28	47.97	9.95
Bilaspur	242	17.59	35.95	24.35
Dantewada*	41	15.56	48.51	7.23
Dhamtari*	209	19.72	46.23	13.24
Durg	329	16.88	33.38	38.15
Jangir-Champa*	342	18.55	37.91	11.03
Jashpur*	127	12.71	49.62	4.60
Kanker*	100	18.84	51.17	4.82
Kawardha*	138	13.86	47.59	7.68
Korba*	153	22.55	33.04	36.27
Koriya*	89	16.91	39.28	29.81
Mahasamund*	180	8.72	40.61	11.35
Raigarh	179	18.68	39.20	13.39
Raipur	231	18.97	33.49	30.42
Rajnandgaon	159	17.70	47.36	18.05
Surguja	125	24.58	43.99	6.96

* *Newly created districts in the Census of India, 2001.*

Appendix III

District Profile of Jharkhand for the Year 2001

	Density	*Decadal Growth Rate (1991-2001)*	*Female Work Participation Rate*	*% Urban to Total Population*
Jharkhand	338	23.19	26.41	22.24
Bokaro*	617	22.11	11.84	45.26
Chatra*	214	29.05	28.18	5.31
Deoghar	470	24.46	23.80	13.72
Dhanbad	1147	22.82	9.47	52.37
Dumka	283	17.31	34.69	6.53
Garhwa*	256	29.05	30.00	4.12
Giridih	387	27.09	21.85	6.43
Godda	497	21.61	28.81	3.53
Gumla	148	16.6	44.93	5.48
Hazaribagh	382	24.02	23.46	23.23
Kodarma*	334	26.32	24.83	17.37
Lohardaga	244	26.14	17.49	12.67
Pakaur*	389	24.34	35.65	5.13
Palamu	241	26.8	27.84	5.96
Paschimi Singhbhum	210	16.35	36.77	16.85
Purbi Singhbhum	561	22.66	20.18	55.03
Ranchi	362	25.72	29.14	35.11
Sahibganj	580	25.89	31.91	10.58

* *Newly created districts in the Census of India, 2001.*

Appendix IV

District Profile of Madhya Pradesh for the Year 2001

	Density	*Decadal Growth Rate (1991-2001)*	*Female Work Participati on Rate*	*% Urban to Total Population*	*Operat-ional Holding*	*Forest Area*
M.P.	196	24.34	33.21	26.46	2.3	8606.4
Balaghat	162	5.85	46.35	12.95	1.4	505.6
Barwani*	199	29.87	43.71	14.61	NA	353.5
Betul	139	18.02	37.05	18.57	2.9	395.6
Bhind	320	17.06	22.87	23.73	2.2	8.9
Bhopal	665	35.91	14.94	80.43	3.3	44.1
Chhatarpur	170	27.33	29.70	21.99	2.3	214.3
Chhindwara	157	17.86	33.02	24.45	2.3	478.9
Damoh	148	20.46	35.80	18.85	2.3	267
Datia	233	21.82	43.23	21.89	2.5	20.8
Dewas	186	26.39	36.42	27.39	3.7	205.7
Dhar	213	27.29	40.49	16.56	3.1	119.4
Dindori*	78	13.23	55.03	4.63	NA	34.8
East Nimar	159	19.31	34.43	26.85	3.2	511.8
Guna	151	27.11	29.15	21.29	2.8	152.1
Gwalior	358	26	14.66	60.23	2.4	109.5
Harda*	142	24.53	36.43	21.32	NA	104.7
Hosangabad	162	22.4	20.53	30.84	3.7	253.2
Indore	633	40.82	19.14	70.17	3	52.2
Jabalpur	413	22.59	22.67	57.05	1.7	73.7
Jhabua	206	23.56	50.69	8.68	2.2	130.9
Katni*	215	20.61	31.30	21.17	NA	100
Mandla	154	14.66	48.31	10.28	2.4	583.2
Mandsaur	214	23.67	40.38	18.64	2.4	36.8
Morena	319	24.09	23.75	21.55	1.8	50.7
Narshimhapur	187	21.88	29.13	15.99	2.5	136.2
Neemuch*	171	21.25	40.14	27.95	NA	68
Panna	120	24.17	35.14	12.63	2	299.4

Contd...

Contd...

Raisen	133	27.8	21.88	18.38	3.6	333.4
Rajgarh	204	26.24	44.84	17.33	2.4	18.8
Ratlam	250	24.97	36.21	30.31	2.7	34.5
Rewa	313	26.84	37.72	16.24	2.1	67
Sagar	197	22.7	30.40	29.22	2.5	288.8
Satna	249	27.52	30.60	20.62	1.8	203.5
Sehore	164	28.22	34.14	17.96	3.5	171.8
Seoni	133	16.49	43.48	10.35	2.6	327.8
Shahdol	158	18.87	34.01	25.27	2.1	304.3
Shajapur	208	24.87	42.01	18.54	3.1	6.8
Sheopur*	85	29.72	31.53	15.83	NA	282.8
Shivpuri	140	27.16	37.30	16.62	2.6	330.3
Sidhi	174	33.28	34.24	14.26	2	438
Tikamgarh	238	27.88	41.27	17.68	1.7	66.6
Ujjain	281	23.63	33.88	38.74	3.6	3.1
Umaria*	127	22.58	33.34	16.11	NA	237.2
Vidisha	165	25.18	21.23	21.43	4.4	105.8
West Nimar	190	27.95	39.49	15.39	3	74.9

** Newly created districts in the Census of India, 2001.*

Notes: 1. Operational Holding in hectare/holding for the year 1995-96, Monthly Review of Madhya Pradesh Economy, CMIE, Regional Monitoring Service, September 2005.

2. Forest area in '000 hectares for the year 1998-99, Monthly Review of Madhya Pradesh Economy, CMIE, Regional Monitoring Service, September 2005.

Appendix V

District Profile of Uttaranchal for the Year 2001

	Density	*Decadal Growth Rate (1991-2001)*	*Female Work Participation Rate*	*% Urban to Total Population*	*Operational Holding*	*Forest Area*
Uttaranchal	159	19.2	27.33	25.67	0.9	3466.2
Almora	205	3.14	47.92	8.64	1.0	327.7
Bageshwar	108	9.21	49.90	3.13	NA	66.2
Chamoli	48	13.51	44.06	13.69	0.9	439.0
Champawat	126	17.56	36.93	15.04	NA	66.0
Dehradun	414	24.71	12.57	52.94	0.9	211.7
Garhwal	129	3.87	36.84	12.89	0.8	444.0
Hardwar	612	26.3	8.79	30.84	0.8	37.5
Nainital	198	32.88	23.88	35.27	0.8	302.5
Pithoragarh	65	10.92	42.58	12.94	0.9	264.4
Rudraprayag	120	13.44	47.15	1.20	NA	127.3
Tehri Garhwal	148	16.15	42.54	9.90	0.9	367.9
Udhamsingh Nagar	424	27.79	13.68	32.62	0.7	101.1
Uttarkashi	37	22.72	43.71	7.77	0.9	710.8

Notes: 1. Operational Holding in hectare/holding for the year 1995-96, Monthly Review of Uttaranchal Economy, CMIE, Regional Monitoring Service, September 2005.

2. Forest area in '000 hectares for the year 1999-2000, Monthly Review of Uttaranchal Economy, CMIE, Regional Monitoring Service, September 2005.

Appendix VI

District Profile of Rajasthan for the Year 2001

	Density	*Decadal Growth Rate (1991-2001)*	*Female Work Participation Rate 2001*	*% Urban to Total Population*	*Operational Holding*	*Forest Area*
Rajasthan	165	28.33	33.49	23.39	4	2645.1
Ajmer	257	26.10	27.97	40.09	2.3	54.7
Alwar	357	30.23	43.88	14.53	1.8	77.8
Banswara	298	29.84	43.92	7.15	1.6	112.8
Baran*	146	26.19	35.60	16.84	NA	215.4
Barmer	69	36.83	41.78	7.40	12.4	30.7
Bharatpur	415	27.05	32.99	19.46	1.8	30.3
Bhilwara	193	26.14	38.61	20.60	2	73.9
Bikaner	61	38.18	27.52	35.54	10.8	81.3
Bundi	167	24.8	40.51	18.65	2.4	141.8
Chittaurgarh	166	21.46	46.23	16.04	2.3	193.7
Churu	114	24.6	38.44	27.87	9.6	6.3
Dausa*	384	32.42	36.12	10.31	NA	24.4
Dhaulpur	324	31.13	33.97	17.96	0.4	27.1
Dungarpur	294	26.58	44.67	7.30	1.4	61.2
Ganganagar	163	27.53	24.96	25.34	7.3	59.3
Hanumangarh*	157	24.34	29.89	20.00	NA	15
Jaipur	471	35.1	22.28	49.36	3.1	81.4
Jaisalmer	13	47.45	29.37	15.03	13.1	23.3
Jalor	136	26.78	46.09	7.59	6	20.1
Jhalawar	190	23.34	39.55	14.25	2.6	121.1
Jhunjhunu	323	20.9	32.61	20.65	2.8	39.6
Jodhpur	126	33.77	27.09	33.85	8.7	7
Karauli*	219	29.96	34.22	14.21	NA	170.8
Kota	301	28.52	19.41	53.46	3	121.7
Nagaur	157	29.33	32.93	17.20	6	18
Pali	147	22.39	30.94	21.47	3.9	80.1

Contd...

Contd...

Rajsamand*	256	19.88	30.02	13.04	NA	24.4
Sawai Madhopur	248	27.44	35.55	19.04	2.1	78.6
Sikar	296	24.11	31.45	20.65	3.1	61
Sirohi	166	30.08	30.11	17.73	2.7	154.8
Tonk	168	24.24	38.27	20.89	3.4	26
Udaipur	196	27.37	31.83	18.62	1.6	411.2

* *Newly created districts in the Census of India, 2001.*

Notes:
1. Operational Holding in hectare/holding for the year 1995-96, Monthly Review of Rajasthan Economy, CMIE, Regional Monitoring Service, October 2005.
2. Forest area in '000 hectares for the year 2000-01, Monthly Review of Rajasthan Economy, CMIE, Regional Monitoring Service, October 2005.

Appendix VII

District Profile of Bihar for the Year 2001

	Density	*Decadal Growth Rate (1991-2001)*	*Female Work Participation Rate*	*% Urban to Total Population*
Bihar	881	28.43	18.84	10.46
Araria	763	31.84	25.82	6.13
Aurangabad	609	30.19	20.15	8.45
Banka*	533	24.47	28.16	3.51
Begusarai	1225	29.11	15.93	4.58
Bhagalpur	943	27.24	21.45	18.67
Bhojpur	937	24.58	12.57	13.93
Buxar*	823	29.03	11.89	9.20
Darbhanga	1446	30.85	14.88	8.11
Gaya	698	30.03	25.47	13.70
Gopalganj	1059	26.11	15.05	6.07
Jamui*	452	32.9	34.78	7.38
Jehanabad	965	28.64	27.91	7.37
Kaimur (Bhabua)*	383	30.64	20.74	3.24
Katihar	783	30.91	23.06	9.12
Khagaria	862	29.32	22.93	5.96
Kishanganj	688	31.5	10.24	9.95
Lakhisarai*	653	23.94	23.28	14.68
Madhepura	854	29.45	37.26	4.45
Madhubani	1021	26.08	20.10	3.48
Munger	802	20.34	13.38	27.94
Muzaffarpur	1181	26.74	12.71	9.30
Nalanda	1007	18.64	26.98	14.92
Nawada	726	33.08	25.77	7.65
Pashchim Champaran	582	30.4	23.57	10.17
Patna	1474	30.17	13.33	41.57
Purba Champaran	993	29.27	14.69	6.37
Purnia	788	35.23	23.32	8.74
Rohtas	636	27.71	13.52	13.33

Contd...

Contd...

Saharsa	894	33.03	28.35	8.30
Samastipur	1169	25.63	15.13	3.64
Saran	1230	26.37	10.37	9.19
Sheikhpura*	763	24.96	24.98	15.47
Sheohar*	1478	36.16	9.77	4.12
Sitamarhi	1169	32.58	11.20	5.71
Siwan	1223	24.78	12.78	5.51
Supaul*	714	29.95	32.99	5.09
Vaishali	1335	26.39	10.91	6.87

* *Newly created districts in the Census of India, 2001.*

Appendix VIII

District Profile of Uttar Pradesh for the Year 2001

	Density	*Decadal Growth Rate (1991-2001)*	*Female Work Participation Rate*	*% Urban to Total Population*	*Operational Holding*[1]	*Forest Area*[2]
U.P.	690	25.8	16.54	20.78	0.9	1688.7
Agra	899	31.27	7.24	43.30	1.3	45.4
Aligarh	820	22.08	13.31	28.90	1.1	2.7
Allahabad	900	26.72	23.05	24.45	0.8	19.9
Ambedkar Nagar*	863	24.31	21.83	8.93	NA	0.4
Auraiya*	586	14.7	14.77	14.32	NA	2.7
Azamgarh	972	26.28	20.22	7.55	0.6	0.1
Baghpat*	881	13	15.14	19.71	NA	1.5
Bahraich	539	29.55	16.13	10.00	0.9	67.7
Ballia	926	21.67	15.44	9.77	0.7	NA
Balrampur*	502	23.08	31.22	8.06	NA	58.9
Banda	345	18.49	30.26	15.87	1.5	5
Barabanki	607	26.4	19.96	9.30	0.7	6.3
Bareilly	878	26.96	9.18	32.93	0.9	0.2
Basti	776	22.69	23.00	5.56	0.7	4.1
Bijnor	687	27.16	7.92	24.31	1.3	43.8
Budaun	594	25.36	7.59	18.15	0.9	6.9
Bulandshahar	669	22.22	28.25	23.15	1.2	7.5
Chandauli*	647	28.63	18.93	10.56	NA	77.4
Chitrakoot*	242	34.33	36.06	9.99	NA	89.2
Deoria	1069	25.03	16.59	9.89	0.6	0.3
Etah	628	24.2	7.90	17.33	0.9	3.1
Etawah	579	21.59	6.32	23.01	1	36.1
Faizabad	892	23.87	27.85	13.46	0.6	3.8
Farrukhabad	720	22.8	7.65	21.75	0.7	1
Fatehpur	556	21.4	25.48	10.30	0.9	6.1
Firozabad	870	33.44	6.88	30.32	1.4	8.6
Gautam Buddha Nagar*	834	35.7	10.90	37.39	NA	0.4
Ghaziabad	2866	47.47	8.62	55.20	1	2.5
Ghazipur	899	26.18	20.80	7.68	0.8	NA

Contd...

Contd...

Gonda	691	25.46	15.77	7.03	0.7	12.9
Gorakhpur	1135	23.44	17.06	19.59	0.7	5.8
Hamirpur	244	17.85	27.38	16.65	1.9	23.5
Hardoi	568	23.67	10.01	11.99	0.9	10.4
Hathras*	726	18.32	10.03	19.80	NA	2.1
Jalaun	319	19.39	20.32	23.41	1.7	25.6
Jaunpur	969	21.67	21.88	7.40	0.5	0.1
Jhansi	347	23.23	23.12	40.79	1.6	33.6
Jyotiba Phule Nagar*	667	29.72	14.13	24.56	NA	11.9
Kannauj*	664	19.58	14.62	16.70	NA	4.8
Kanpur Dehat	517	21.55	14.69	6.89	1	10.2
Kanpur Nagar	1321	27.17	9.71	67.12	0.8	1.6
Kaushambi*	726	26.73	30.07	7.10	NA	1
Kheri	418	32.28	8.88	10.77	1	164
Kushinagar*	996	28.17	22.52	4.58	NA	0.8
Lalitpur	194	29.98	34.99	14.52	2	76.2
Lucknow	1443	33.25	10.44	63.63	0.8	21
Maharajganj	736	29.27	29.92	5.09	0.7	45.5
Mahoba*	246	21.8	32.70	21.86	NA	15.4
Mainpuri	579	21.5	6.30	14.60	0.8	2.2
Mathura	621	26.95	24.46	28.30	1.6	1.5
Mau	1082	27.91	22.52	19.44	0.7	0.6
Meerut	1157	24.16	10.73	48.44	1.2	26.8
Mirzapur	468	27.62	20.62	13.54	1.1	107.9
Moradabad	1025	26.45	11.67	30.54	1.1	0.2
Muzaffarnagar	884	24.61	14.86	25.51	1.2	17.4
Pilibhit	470	28.11	5.56	17.88	1.2	80.2
Pratapgarh	735	23.36	24.21	5.29	0.5	0.6
Rae Bareli	623	23.66	22.71	9.54	0.7	4.9
Rampur	813	27.98	6.90	24.97	1.1	6.6
Saharanpur	785	23.35	6.34	25.81	1.3	33.2
Sant Kabir Nagar*	863	23.64	23.65	7.08	NA	1.5
Sant Ravidas Nagar Bhadohi*	1334	25.47	15.03	12.82	NA	0
Shahjahanpur	557	28.28	4.66	20.63	1	10.5
Shravasti*	479	27.3	25.95	2.84	NA	34.4

Contd...

Contd...

Siddharthnagar	705	26.78	27.95	3.81	0.7	3.6
Sitapur	630	26.58	8.51	11.95	0.8	5.8
Sonbhadra	216	36.13	25.03	18.82	1.3	363.4
Sultanpur	725	24.2	19.33	4.74	0.6	2.1
Unnao	592	22.72	16.79	15.24	0.8	16.7
Varanasi	2045	25.51	16.47	40.16	0.5	0.7

* *Newly created districts in the Census of India, 2001.*

Notes: 1. Operational Holding in hectare/holding for the year 1990-91, Monthly Review of Uttar Pradesh Economy, CMIE, Regional Monitoring Service, September 2005.

2. Forest area in '000 hectares for the year 2001-02, Monthly Review of Uttar Pradesh Economy, CMIE, Regional Monitoring Service, September 2005.

NA – Not Available.

Notes

1. An operational landholding was defined as "all land which is used wholly or partly for agricultural production and is operated as one technical unit by one person alone or with others without regard to the title, legal form size or location".
2. Weight-for-age is a composite measure that takes into account both chronic and acute undernutrition. Children who are more than two standard deviations below the reference median on this index are considered to be 'underweight'.
3. Height-for-age index measures linear growth retardation. Children who are more than two standard deviations below the median of the reference population in terms of height-for-age are considered short for their age or 'stunted'.
4. Weight-for-height index examines body mass in relation to body length. Children who are more than two standard deviations below the median of the reference population in terms of weight-for-height are considered too thin or 'wasted'.

References

Acharya, Sanghmitra (1997): "Women in the Context of Environment and Sustainable Development", *IASSI Quarterly*, Vol. 16 (3 & 4), pp. 275-291.

Agarwal, Anil (1985a): "The Environment and Rural Women", *Indian Express Magazine*, June 2.

Agarwal, Bina (1981): "Agricultural Modernization and Third World Women, Pointers from the Literature and on Empirical Analysis", Geneva, Switzerland: ILO, Rural Employment Policy Research Paper.

Basu, A. (1991): *Culture, the Status of Women and Demographic Behaviour*, New York: Oxford University Press.

Batliwala, Srilata (1983): "Women and Cooking Energy", Paper presented at a seminar on "Cooking Energy Systems: Problems and Opportunities", quoted in *The State of India's Environment*, op cit.

Bose, A. (2001): *Population of India, 2001*, New Delhi: B.R. Publishing Corporation.

Centre for Monitoring Indian Economy (2000). Profiles of Districts, October 2000, Mumbai: Economic Intelligence Service.

Centre for Science and Environment (1985): "The State of India's Environment – The Second Citizens Report", New Delhi.

Directorate of Economics and Statistics (2004): "Agricultural Statistics at a Glance", August, New Delhi.

Government of India (2001): "State of Forest Report 2001", Ministry of Environment and Forest, GOI, Dehradun.

Government of India (2003): "Compendium of Environment Statistics, India, 2003", Central Statistical Organization, Ministry of Statistics and Programme Implementation, GOI, New Delhi: Akashdeep Printers.

Gupta, Kamla (1997): "Population Growth, Environmental Degradation and its Consequences", in Gupta, Kamla and Arvind Pandey (eds.), *Population and Development in Uttar Pradesh*, Delhi: B.R. Publishing Corporation.

Gupta, P.K. and B.C. Joshi (1987): "Fragile Ecology of Himalayas," in *Environmental Management in India*, Vol. 1, R.K. Sapru (ed.), New Delhi: Ashish Publishing House, pp. 221-229.

International Institute for Population Sciences (IIPS) and ORC Macro 2000 (1998): *National Family Health Survey (NFHS-II)*, Mumbai, India.

Jejeebhoy, S.J. (1990): "Women's Roles: Health and Reproductive Behaviour", in Satia, J.K., and Shireen J. Jejeebhoy (eds.), *Strategies for Fertility Decline in the Four Large North Indian States: Bihar, Madhya Pradesh, Rajasthan and Uttar Pradesh*, Oxford University Press.

Jodha, N.S. (1985): "Population Growth and the Decline of Common Property Resources in Rajasthan, India," *Population and Development Review*, Vol. 11 (2), pp. 247-264.

Ministry of Agriculture (1981): "Agricultural Situation in India", in *Agricultural Statistics at a Glance,* October, New Delhi.

Monthly Review of Uttar Pradesh Economy, (2005): *CMIE,* Regional Monitoring Service, Lucknow, September 2005.

Monthly Review of Uttaranchal Economy, (2005): *CMIE,* Regional Monitoring Service, Dehradun, September 2005.

Monthly Review of Madhya Pradesh Economy, (2005): *CMIE,* Regional Monitoring Service, Bhopal, September 2005.

Monthly Review of Rajasthan Economy, (2005): *CMIE,* Regional Monitoring Service, Jaipur, October 2005.

Nagbrahman, D. and Shreekant Sambrani (1983): "Women's Drudgery in Firewood Collection", *Economic and Political Weekly,* Vol. XVIII, No. 1 and 2, January 1-8.

Registrar General of India (2001): "Provisional Population Totals, Series-1, Paper 1 of 2001, Supplement District Totals", New Delhi: GOI.

Swaminathan, M. (1985): "A Study of Energy Use Patterns of Two Villages of Garhwal Himalayas", in *The State of India's Environment,* op. cit., pp. 174.

Tabah, L. (1992): "Population Prospects with Special Reference to Environment, Population and Environment", Inter-Centre Cooperative Research Programme, CICRED, Paris.

World Bank (1991): "Gender and Poverty in India: A World Bank Country Study", The World Bank: Washington D.C.

3 Vanishing Common Property Resources and their Environmental Impact in Rural Areas: Examples from Two Districts of Andhra Pradesh

M. Ramakrishna Reddy

The true capacity of the earth to sustain the ever growing population appears to have been really exhausted despite the neo-Malthusian arguments that the human ingenuity leading to great strides in the production of food and other needs of the populations. In developing countries the question of demographic bonus arising out of the growing population does not seem to be attractive at all or at the least it could be a wishful thought that never proved itself. However, with the passage of time, the mounting pressure of population growth is affecting the natural resources to the extent of having a severely undermining of the delicate environment in which the very populations exist. The fragile elasticity between the infinitely expandable population and the finite natural resources reminds that the forebodings of the Malthusian theory have not been totally silenced by the alternatives suggested ever since Malthus essayed the debate.

Observations in India and other developing countries point to the fact that human ingenuity and technology have not really made any difference in the face of growing population size. The technological marvels of the modern times – the green, the white, the blue revolutions and so on have not halted the growing inroads into the natural resources base of the communities destabilizing the very foundation of existence especially in the rural areas. In fact, the very vital sphere of life on this planet – the environment – is absorbing the shock of such an impact. Nevertheless, such a forbearance of the nature must also have a limit, about which we know or have understood very little. The over- all picture of race between the increasing size of population and the shrinking natural resources is becoming glaring with the passage of time and the population emerging as the winner. However, this losing race is reflected in some of the natural assets like common property resources getting wiped out slowly, but certainly, as observed in the two districts of Andhra Pradesh over the past one decade. These observations do reveal the very close nexus between the growth of population and gradual, yet, certain environmental degradation. The intent of this paper is to delve into one minor facet of the environmental degradation that was observed in two districts of Andhra Pradesh and to examine the nexus between the population growth and the environmental degradation.

Common Property Resources vis-à-vis Rural Living

In the recent past the common property resources which abound the rural landscape, were a set of dependable resources, which provide a variety of economic and environmental security for the whole of the community, especially, in the developing countries. A substantial part of the rural economy of the past was dependent on these valuable resources like common grazing grounds, community animal stalls, animal watering points, waste dumps, wood lots, water bodies and streams, and marginal unreserved forests. Promoted with common weal, their extent and integrity was respected, and managed with concern under notional ownership of the entire community the marginal resources were intended for the welfare of the community as a whole (Jodha, 1985). Access to such resources was free and least competed for private acquisition. The slow population growth of the era did not exert undue pressure on such resources. These open natural resources have provided cushioning effects to the rural subsistence economy for generations. These resources were more critical in subsistence economic zones of the rural areas as they offered unrestricted access to several economic benefits to the communities and thus sparing the resources with superior value and utility from being used for marginal productive activity.

This paper is particularly focussed on the scrub forests as a common property resource that stood on the degraded marginal lands in the vicinity of the villages providing marginal economic support for the communities. While providing economic support these resources indirectly promoted well-being of the people by providing environmental security by preventing soil erosion, runoff of rainwater and providing vegetative cover to the marginal lands for a larger part of the year. This is particularly important in certain regions of the country facing the bleak prospects of desertification.

Common Property Resources: The Rustic Wisdom for Survival

Since ages, the rural communities in India and other developing countries have taken into consideration of common weal of the people and have provided for open spaces for which we long for now. These properties were notionally owned and operated by the communities themselves. In their wisdom these communities have earmarked open free spaces to serve the common utility of the community with free access to these properties. In the past, these properties virtually met all the social and economic needs of the people. The extent of such properties, which existed for centuries, has started dwindling since 1960s and some of them have almost vanished now. The collective rustic wisdom of the past seems to have been forced to retract into the background, yielding way for gradual encroachments thus common properties are well on their way to extinction. The scrub forests that existed about a decade ago have gradually been converted into crop-lands leaving the villages with no grazing grounds is the subject matter of this paper.

It does not seem to puzzle anybody about the plausible reasons for the reduction or loss of such common property resources. Because, the growth and size of the population is directly proportionate to the extent of the loss of these common property resources. The impact of the population pressure is very obvious on these properties.

The Case of Two Districts of Andhra Pradesh

Anantapur and Kurnool districts are situated in the south-western part of the Andhra Pradesh bordering Karnataka. These two districts are a part of the semi-arid tropical Rayalaseema, and Anantapur district is considered to be in the process of desertification and consequently receives the second lowest rainfall in the country after Jaisalmer in Thar Desert of Rajasthan. Kurnool district is endowed with thick Nalla Mala forest cover on its eastern side but the western and southern parts of the district are contiguous with Anantapur district and topography is similar to Anantapur district.

Anantapur district has a surface area of 19,130 square kilometres with a population of 3,640,478 as enumerated during the 2001 census and a population density of 190 per sq km. Kurnool district has a surface area of 17,658 square kilometres with a population of 3,529,494 and a population density of 199 per sq km, as enumerated by the 2001 census (Table 1). The population density in these two districts has shown a steady increase since 1961 and has doubled itself in four decades, from about 100 per square kilometre in 1961 to about 200 per square kilometre in 2001. This doubling of the population within the rigid inelastic limits of land area has forced people to look for other available natural resources for sustenance and that seems to have triggered the process of encroachment of common property resources in these two districts.

Table 1

Population Size, Growth, and Density in Anantapur and Kurnool Districts of Andhra Pradesh

District	*Year*				
	1961	*1971*	*1981*	*1991*	*2001*
Anantapur					
Population	1,767,464	2,115,321	254,821	3,183,814	3,640,478
Decadal Growth Rate	19.3	19.70	20.46	24.95	14.31
Area in sq km	19,130.00				
Population Density	92	111	133	200	190
Kurnool					
Population	1,908,740	1,982,090	2,407,299	2,973,024	3,512,266
Decadal Growth Rate	18.01	21.45	23.50	26.20	13.00
Area in Sq Km	17,658.00				
Population Density	107	105	136	168	199
Andhra Pradesh					
Population	35,983,447	43,502,708	53,459,673	66,508,008	76,210,067
Decadal Growth Rate	15.65	20.90	23.10	24.20	13.86
Area in sq km	275,045.00				
Population Density	131	157	195	242	275.00

Sources: Census of India (various years).

The decennial growth rate in these two districts presents a remarkable picture of the continuous increase in size of the population since 1961. The growth in the size of population has been more among the poor, the landless and the marginal populations. This segment of population is the one which does not generally own economic assets. With the growing poverty and felt need for owning economic assets for sustainable support seems to have forced certain sections of this population to encroach on the marginal common property resources in the rural landscape. Secondly, while the available arable land in these districts remained constant and inalienable, the additional increase in population density has forced people to look for resources of subsistence within the severely limited primary resources leading to encroachment on the marginal resources, which have hitherto served as the common property resources.

It is interesting to note that the beginning of the population growth in India almost coincided with the country gaining independence leading to the establishment of the constitutional democracy with adult franchise, i.e., universal 'voting' rights and followed by land reforms. The statecraft in India is based on the votes cast during several political campaigns at regular intervals. The quest for political power has created a situation wherein trade-offs of different kinds have come to exist. One such trade-off is the tacit support lent by local political pressure groups to certain sections of people to usurp common property assets in and around the villages. It is a working arrangement commonly observed in these two districts.

Types of Common Property Resources

The common property resources described in this paper are large stretches of scrub forests and wastelands standing on the marginal degraded land in the vicinity of villages. Some of these stretches were plain while many were undulation interspersed with rocky boulders of varying sizes. The quality of soil is poor and the vegetation and rocks prevented soil erosion over years. The quality of soil on these stretches of land varied from shallow layers of red compact loams, to sandy loams of poor quality and occasionally shallow layers of black cotton soils. The poor quality of soil attracted least interest from the villagers in the past and also since they served as the village grazing lands, waste dumps, and thrashing yards, the villagers were not interested in annexing these pieces of land.

The scrub forests on these lands used to be a thin bushy vegetation, covered by a large variety of cacti and annual varieties of grasses and these stretches of land served as the village pastures for cattle and sheep. There were no trees on these lands but partially covered by woody bushes which supplemented the fuel wood needs of the villages. These grazing grounds looked green during the rainy season and dry during summer months. These were the only pieces of free land and space available around the village for the common access and use, which have since vanished leaving behind a large livestock without free grazing grounds. Secondly, the attempt to cultivate these lands has created another vicious cycle by disturbing the structure of the existing soil surface, leading to soil erosion and land degradation due to runoff rainwater carrying away the precious little topsoil from these already poor lands.

These scrub forests were surrounded by crop-lands with varying degree of soil fertility. It is again monsoon dependent subsistence agriculture, of the semi-arid tropical type. Common crops like groundnut, pearl millet, sorghum, pigeon pea, chick pea, and other varieties of pulses are grown depending upon this type of soil. The area of individual holdings of the cultivated land, though varying, was not very extensive.

Objective

The objective of this paper is two-fold:

1. to highlight the growing pressure of population on common property resources in the observation districts, and
2. to focus on the economic outcome and environmental degradation consequent to population pressure on these common property resources with a suggestion for further research to understand the environmental impact of the vanishing scrub forests in these districts and its long-term impact.

Methodology and Information Sources

This is a preliminary enquiry and an observational study spread over a period of twelve years. During this period, the common property resources, viz., scrub forests in the vicinity of the villages were observed keenly noting the extent of encroachment every season and through all the years till the entire extent was converted into cultivable pieces of land. Although the pattern of encroachment is found all over the district without exception, for the purpose of this paper several sites which lay adjacent to the major roads in the districts and are quite familiar by sight for observation have been selected. The common property resources described in these districts are two sites in

Anantapur and two sites in Kurnool district. The sites in Anantapur are Sitaramapalli scrub forest and Thimmapuram hillock range on the National Highway 4. In Kurnool district, the stretches of scrub forests on the state highway around the Adoni cross near Alur and large stretches along the National Highway No. 4 between Veldurthy and Dhronachalm.

The author has travelled all over the districts of Anantapur and Kurnool, since January 1994, as part of the family planning service provision in camps. The travel schedule was monthly, covering the entire district including the interior areas and the observations were made during such visits. The main purpose of this paper is to create awareness about the vanishing scrub forests in the vicinities of the villages which served as common property resources.

The observation was made about the dwindling extent of the scrub forest and the process of conversion of it into pieces of crop-lands for growing the common crops of the area. The observations are presented in a descriptive form. Due to the difficulties in collecting data on encroachments, estimations or statistical analysis has not been attempted.

The Background of Encroachments

The existing political climate in these districts has created a situation which favours and actively supports encroachments on the common property resources. Taking the advantage of the prevailing political situation people encroach upon the common property resources around the villages for their individual benefit, thus depriving the hitherto available access and use of these properties by the entire village. Though not patronized directly, tacit political support flows from the local leadership who are the building blocks of the political parties who eulogise the common man – the voter, as the real masters of the political rulers.

These villages face several peculiar situations socially and politically, which actively facilitate encroachment on common property resources. It is common knowledge that the village populations in the recent decades are segmented into groups either on the criteria of socio-economic status or group dominance or political affiliation. Marginal groups of people like agricultural labourers belonging to all caste groups, more so, the schedule caste and schedule tribes, dominant caste groups of the villages, and groups with political backing indulge in encroachment on these lands. Oppressed over a long period these groups have become very vocal and over the years have come to understand that they can assert and accordingly agitate at the slightest instance. They are never restrained either by the well-meaning villagers, public administrators, or the local political leadership. Instead, as a matter of fact they are encouraged to indulge in encroaching on the common property resources with a view to protect and expand the ranks of voters as a quid pro quo.

The Modus of Encroachment

The encroachers locate a piece of pastoral land on the lot of common property resources of the village and set out cultivating the piece during the beginning of the rainy season. The extent and boundaries of the piece of land is left vague so that at least in the beginning, the encroachment does not become very obvious for the onlookers, yet has the potential for expanding and settling down upon such land permanently. Initially, the existing vegetation is cleared and land is prepared for cropping in, with common crops grown in the area. The intervening rocky boulders or other barriers are not touched in the early stages. This pattern almost resembles the shifting cultivation practised by tribal populations, elsewhere. This is an initial effort to establish presence on the land sans definitive right to the piece of land.

In first cropping season, due to the effect of tilling the virgin land and some kind of added fertilization (most often chemical fertilizers) the crop appears attractive but in patches interspersed with rocks and other natural barriers like deeply excavated pits or natural storm water drains. Rain water management practices like peripheral and internal bunds are not laid. These lands being undulating with uneven surface have larger potential for erosion and runoff of the rainwater carrying the loosened scarce topsoil. The crop output would be insignificant because of patchy cultivation with inorganic chemical inputs, and expectedly, yield would be varying within the given piece of land. Often noxious pesticides are used to protect the crops on these lands which tend to build up due to scarce water available to leech these chemicals over a period of time.

The immediate post-harvest period appears to be some kind of encouragement for the encroachers as there were no overt objections for the cultivation on the marginal lands and the piece of land wears a cultivated look with ownership posed along. The family gets into act to develop the land further. The rocks and boulders are removed

Photo 1

Encroached scrub forest which was grazing ground earlier for the cattle and sheep in the village Thimmarpuram in Anantapur District, Andhra Pradesh

Photo 2

Cultivated hill slope near Thimmarpuram

Photo 3
Cropped slope of the hill near Thimmarpuram

Photo 4
Cropping parches of land in between rocks and boulders at Adoni Cross, Alur in Kurnool district

Photo 5

Groundnut crop on the encroached land near Dhronachalm in Kurnool District, Andhra Pradesh

Photo 6

Groundnut crop raised on encroached scrub forest near Veldurthy in Kurnool district, Andhra Pradesh

assiduously sometimes piling up the debris as border hedges. The task of clearing the land is difficult and the labour is contributed by the family over several weeks. This process of clearing the field could take two to three or more seasons depending upon the extent and size of the rocks. Also, it appears to be a deliberate act of delaying, so that the piece of wasteland does not appear to be crop-land overnight to attract attention of revenue officials and others in the village. Some of the larger boulders are left alone as their removal would be difficult and expensive (photographs 1-4).

The ownership rights are established gradually with application to the local revenue department for grant of 'government' land. The only evidence required, if at all necessary, is a differently coloured ration card to prove the *locus standi* of the applicant and that the family is either poor or below the poverty line (BPL). It might take a long time for getting the rights assigned in the normal course, but the pace could be hastened with political backing and political process like elections of any kind passing by. The public will never raise objections against the encroachment, or the local political lobbyists will brush aside such objections as these families form strong vote banks for a reasonably long period. Jodha and Gulati have found similar situation in the villages of Rajasthan and Uttar Pradesh in relation to the common property resources (Jodha, 1985; Gulati and Chopra, 1994).

The second pattern of encroachment is to attempt cultivation of slopes of hills and hillocks in the district, which is especially prevalent in Anantapur district. This is a notorious practice which would further soil erosion and land degradation at a very fast pace, which has not been realized. The encroachment is up to the peak of the hills. The slopes are cleared of the scrub forest cover and the land is prepared in the pattern of terrace cultivation with several tiers. Most often the gradient of the land so prepared would be nearly 45 degrees. The cultivation practices used by the encroachers would further soil degradation at a faster pace. The soil erosion resulting from rainwater runoff would be greatest in this case because of the gradient of the land and the dynamics of the rainwater (photographs 5 and 6).

Ecological and Environmental Impact of the Loss of Scrub Forest and Wastelands

The common resource properties are basically situated on the marginal and often degraded stretches of land. The soil cover of these lands is typically shallow to very shallow which is protected by the existing thin cover of annual grasses, sparse vegetation and strong rocks. The deeper layers of soil are poor in quality to support any cropping on them. Being large stretches of land used for grazing these lands are hosts for a variety of micro-ecosystems that support the environmental diversity within their limited extent. The flora though not very rich, has managed to keep the ecosystem intact while providing grazing opportunities for the cattle and sheep of the village for a large part of the year. The organic matter generated in the form of dung and other animal droppings along with the shed foliage forms, the main source of humus, which helps to retain the rain water, supplement soil fertility, and prevent runoff. The rocks and boulders act as natural barriers for large-scale soil erosion.

The desire to acquire a piece of land from out of the already shrinking common property resources of the village contributes to the ever expanding ecological and environmental degradation around the villages across the country. It is particularly critical in the villages of the observation districts. Clearing and preparing the land for agricultural practices leads to disturbances in the existing micro-ecosystem, loosening the soil surface which is prone for easy runoff washing away the precious yet meagre top soil, besides loss of vegetation. The excavation of the interspersed rocks on shallow soil surfaces disturbs the composition of the soil surface. Soil scientist hold the view that it takes more than 2000 years to form a good soil surface suitable for cultivation purpose and any disturbances would be counter productive (Krishna, 2005)

The clearing and tilling of these lands without proper contour bunds would easily promote runoff of the rainwater during the precipitations. The runoff leads to erosion of the loosened, shallow topsoil and repeated gushes of runoffs carry with them the eroded topsoil. The rainwater, which was retained in these lands earlier, now finds it way to the streams eventually ending up in village tanks and other water bodies. The topsoil carried in the runoff would aggregate in the form of silt in the water bodies. The extent of runoff is directly proportionate to the extent of land surface thus disturbed, and so would be the silting of the water bodies, thereby reducing their water holding capacity. The runoff also carries with it the residues of chemical fertilizers and pesticides used in cultivating these lands which would affect the quality of water and the aquatic life in the water bodies.

The accumulation of silt in the water bodies reduces the holding capacity of the water bodies over a period of time, thereby decreasing the water resources both for groundwater recharging and actual usage. The washed off pesticide and fertilizer residues contribute to the quality of water in the water bodies and seriously affect the aquatic life in the water bodies.

Living without CPRs in Anantapur and Kurnool Districts

The population pressure in the two observation districts is quite high and majority of the population is dependent on agriculture, animal husbandry, dairy farming, and sheep and goat rearing. With the constant shrinking of the village grazing grounds and wastelands, people find it difficult to stock the cattle and sheep. Although the crop-lands are available for grazing cattle and sheep in the post-harvest period it is limited to the dry periods of the year. The advantage of grazing animals during the rainy season on the village grazing grounds is lost forever. This would certainly affect the already declining cattle population affecting the fragile rural economy, and a portion of definite nutritional security that was readily available in the form of milk, meat, and manure for crops is lost for ever.

The wastelands around the villages were used as waste dumps, threshing grounds and semi-permanent, open-cast cattle and sheep stalls. Loss of the common properties resources has not only dislocated the other subsidiary agricultural operations around the year but also has forced people to look for alternatives. In the process, asphalted stretches of highways and major roads are encroached periodically for subsidiary agricultural operations like threshing and drying of agricultural produce. This poses hazard for the road users and several road traffic accidents have occurred in the recent past. The harvest season heralds road blockades in these two districts due to post-harvest agricultural operations on the roads. The waste dumps besides conserving the organic agricultural and animal waste products serve as major source of organic manure for the land. But due to encroachment on the CPRs, a good deal of useful agricultural waste is scattered haphazardly resulting in loss of useful recyclable waste in addition to severe public health hazards.

Recommendations

This being an observational study and explorative in nature, the study could only point to the existing problem which seems to have not been subjected to systematic study. Since Anantapur district is already considered to be in the process of desertification the over-exploitation of even the marginal land resources needs to be considered seriously as it might promote the advancement towards speedier process of desertification. Therefore, there is an urgent need for in-depth study on this phenomenon of encroachment of the marginal lands in the vicinity of the villages and suggest means to halt the process further and to contain the damage already done. The extent of encroachment needs to be surveyed, mapped and the pattern of land use and its outcome such as soil erosion and silting of the water bodies need to be assessed. With the ease of access to remote sensing facilities it might be easy to locate these marginal lands for studying the spatial and temporal dimensions of current situation and future land use patterns. In the background of severe water scarcity in these districts in the past couple of years and with the felt need for groundwater conservation it appears necessary to study the pattern of land use, the current status of watershed management and land reclamation to develop comprehensive approaches to prevent further degradation of the marginal land assets.

Conclusions

It has to be realized that the size of the population and its incessant growth have been far beyond the physiological capacity of the arable land resources. In its search for sustainable alternatives, the growing population in these two districts has set itself on the easily available but marginal land assets of the community, which served as common property resources for ages. The common property resources, which were administered under the notional ownership of the community, were easy to encroach upon and the prevailing social and political environment in these districts and the vested interests discreetly support and encourage such a move unmindful of the ultimate outcome that befall the communities concerned. The environmental impact of encroachment of these marginal lands is enormous as the available free space and the much limited green cover is eliminated for good. These marginal lands get further degraded due to soil erosion, secondary to tilling and rainwater runoff. The runoff actively promotes silting of water bodies and contaminates them with the chemical and pesticide residues, harming the people as well as the aquatic life in such water bodies. The reduced water holding capacity leads to floods at the times of heavy precipitations in rainy seasons affecting even the water storage capacity of the reservoirs down stream due to extended silting. Irretrievable environmental degradation was apparent in the stretches of scrub forests observed in the two districts of Andhra Pradesh as a consequence of increase in the size of the population in these two districts. The communities in the vicinity have lost even the minor economic benefit arising out of these scrub forests.

References and Further Readings

Dasgupta, P. (1998): "The Economics of Poverty in Poor Countries", *Scandinavian Journal of Economics.* 100(1):41-68.

Gulati, S.C. and K. Chopra (1994): " Population Redistribution, Environmental Degradation and Landuse Patterns: A District Level Study of Linkages in Arid and Semi-arid Zones of India: *Demography India.* 1994 Jan-Dec; 23(1-2): 1-14.

Gupta, K. (1997): "Population Growth, Environmental Degradation and its Consequences for Women's Development in Uttar Pradesh". In: *Population and Development in Uttar Pradesh*, edited by Kamla Gupta and Arvind Pandey. New Delhi, India, B.R. Publishing Corporation, pp. 415-29.

Jodha, N.S. (1985): "Population Growth and the Decline of Common Property Resources in Rajasthan, India", *Population and Development Review* 11(2). New York, The Population Council, pp. 247-264.

– (1990): "Depletion of Common Property Resources in India: Micro-level Evidence", In: *Rural Development and Population: Institutions and Policy*, edited by Geoffrey McNicoll and Mead Cain, New York, The Population Council, pp. 261-283.

Kadekodi, G.K. (1992): "Paradigms of Sustainable Development", *Development.* 1992; (3):72-6.

Krishna, Vishal (2005): "Lure of Cash Turns Farmland into Sand Mine: Bangalore", *The New Indian Express*, September 19, 2005.

Kurien, J. (1992): "Ruining the Commons and Responses of the Commoners: Coastal Overfishing and Fishworkers' Actions in Kerala State, India", In: *Grassroots Environmental Action: People's Participation in Sustainable Development*, edited by Dharam Ghai and Jessica M. Vivian. London, England, Routledge, pp. 221-58.

Lee, R.D. (1990). "Comment: The Second Tragedy of the Commons", *Population and Development Review.* 16 Suppl: 315-22.

Repetto, R. (1994): "The 'Second India' Revisited: Population, Poverty, and Environmental Stress over Two Decades". *Wri Publications Brief.* August: 1-4.

Stern, P.C. (1993): "A Second Environmental Science: Human-Environment Interactions". *Science.* 25;260:1897-9.

4 Concern for Environmental Degradation in India's Planning: A Review

Sancheeta Ghosh

India's environmental problems are gaining global significance because of the rapid and aggressive speed of urbanization and lack of infrastructure. Increasing urbanization, industrialization, deforestation and transportation, the second cousins of economic development put tremendous pressure on natural resources, and therefore, there is a pressing need to strike a balance between developmental planning and urgency to safeguard the environment.

As a means of self-preservation, people in India, as in other parts of the world, have been denuding the forests, digging up the land, mining the mountains and farming the seas and so forth. While they were inflicting serious damages to the ecosystem, the regenerative power of the nature, were generally able to heal the environmental damage (Fernandes, 1996).

Today, people are destroying faster than nature can replenish, because of our numerical strength of ten million people, and our insatiable desire of producing more and more effective tools for conquering nature. We in India have been polluting the water and air and degrading land faster than nature can purify them.

In India, the environmental problems are caused because of underdevelopment. Millions of our people still continue to live far below the minimum levels required for a decent human existence, deprived of adequate food, clothing, shelter, education, health and sanitation. Poverty and lack of alternatives are the forces which drive rural people in India to the burning of forests, tilling of marginal lands, the over dependence on finding grazing land for cattle and the over-cutting of trees for fuel. We can see the other side of the coin also. Over-technological and economic development leads people to live in poor environmental condition. Increasing number of vehicles, construction of industries and dams are the causes of environmental degradation in a country like India. Both urban and rural areas in India are overwhelmed by large numbers of people, who no doubt are looking for opportunity to join the process of development. Green revolution has already forced large numbers of rural people to migrate to cities. In India, we are faced with the size of the population and its uneven distribution; poor housing conditions, slums, squatter settlements, inadequate water and sanitation (Fernandes, 1996). The man-made environment of our cities is undergoing violent changes to the extent that authorities seem powerless in grappling with the situation.

Present Situation of Environmental Degradation

India is facing an alarming situation in environmental condition in the present day. Indeed the major environmental problems threatening our society are frightful in their range and severity (Pal, 1999). India ranks the sixth largest and second fastest growing producer of Green House Gases (GHGs) in the world. Three of India's largest cities are considered among world's ten most polluted cities. Nearly 12 years since the disastrous Union Carbide Chemical leak in Bhopal and after five years of economic resurgence, environmental awareness is high: titled GREEN–India (Growth with Resource Enhancement of Environment and Nature). A report by TERI (Tata Energy Research Institute) focusing on the state of the Indian natural resources and environmental pollution was released to mark the 50th anniversary of India's independence. According to the study, India is losing at least 10 per cent of its natural income due to environmental degradation. According to the report, the availability of fresh water declined by two-thirds. The water requirement of major water consuming industries such as agro-based, refineries, petrochemicals, fertilizers has grown 40 times but these are not yet treating the huge waste water generated. Indoors and outdoor air pollution result in almost 2.5 million premature deaths. The total sewage generation from the urban centres has grown six times in the last 50 years (TERI, 2003). It is evidenced from the Table 1 that the ambient air quality levels in some of the cities of India, are of growing concern.

Table 1

Ambient Air Quality Levels in Highly Polluted Cities in India, 1998

(μ g/m3)

	Sulphur Dioxide	*Oxides of Nitrogen*	*SPM (Suspended Particulate Matter)*
Anpara	117	121.7	513
Pune	100.4	119.2	749
Haldia	87.8	115.4	395
Jamshedpur	74.1	93.9	365
Bangalore	40	44	259
Hyderabad	38.9	84.9	776
Bhopal	17.5	26.3	384

Source: CPCB, 2000. National Ambient Air Quality Monitoring Series, Ministry of Environment and Forests, Government of India.

Concern for Environmental Degradation in India's Planning

More than two decades ago the United Nations' conference on the 'Human Environment' in Stockholm (1972) drew attention to the government and people of world to the increasing evidence that our activities were producing deleterious effects on the natural and man-made environment, and creating serious risks for the survival and well-being of people themselves and also to provide the basis for international cooperation, to meet this new challenge of the 20th century.

India is the first country, which has provided for the protection and improvement for the environment in its Constitution and has taken several steps in planning and policies to overcome the environmental problems. If we go to back, we can have the idea that there is an evolutionary process in India's planning from the beginning.

The First and Second Five-Year Plans

After the formal independence the Planning Commission of India prepared the first documented plan in 1952 under the chairmanship of the then Prime Minister of India. The main objective of this plan was to raise the living standard of people and to increase the national income and per capita income by 20 per cent and 17 per cent respectively and open revenues for all people. But, in the First and Second Plans there were no concrete steps for environmental degradation. Only few scattered forest and soil conservation policies were undertaken. The

Forest Policy Revolution Act of 1952, suggested maintaining one-third of its total land area under forests.

The Third Five-Year Plan

In the Third Five-Year Plan, it was proposed not only to intensify some of the programmes initiated under the first and second plans, but also to put special emphasize on more intensive forest and soil conservation programmes. Steps were taken both for utilization of forest resources as well as to protect the forests. Different afforestation programmes and river valley projects were in focus.

Fourth and Fifth Five-Year Plans

The Fourth Five-Year Plan showed no new dimension in environmental concern. Only some few forests policies and programmes were merged with animal husbandry and fisheries section. The Fifth Plan also goes in the same direction. There were no such new areas. Only some few forestry programmes had taken place. But, in this plan, emphasis was given to improve urban environmental condition, basically in the slum areas.

The Sixth Five-Year Plan

Concern of environmental problem made its first footsteps in the Sixth Five-Year Plan. For the first time, in India's planning, there was separate provision for environmental degradation. Steps were taken for water pollution, air, noise and land pollution separately. Though plans and programmes in the field of soil conservation and public health forests and wildlife protection, industrial hygiene, etc., had been in existence in India for many decades, but the first formal recognition of the need for integrated environmental planning was made when the Government of India constituted the National Committee on Environmental Planning and Coordination (NCEPC) in 1972.

The Seventh Five-Year Plan

The Seventh Plan was the continuation of previous plan with some new improvements. The basic approach to the Seventh Plan was sustainable development in harmony with environment. With the realization that poverty and the state of underdevelopment led to many of the environment problems that confronted with nature, hence the understanding that development is the way of environmental management. However, another way of environmental problems has arisen as unintended side effects of the very attempts of development. The plan was made in the following framework:

Pollution Monitoring and Control

Environmental pollution is a serious and growing hazard in India. Its impact on human health and well-being is both direct (inhalation of polluted air or intake of contaminated water) and indirect (loss of soil fertility, death of aquatic life). The Central Board for the Prevention and Control of Water Pollution spearheads the effort such as assessment and control of coastal pollution and strengthening the National River Water Quality Monitoring. A major programme on prevention of pollution of Ganga was planned.

Environmental Impact Assessment (EIA)

It was proposed to induce socio-economic industries and department as well as different agencies whose projects impinge on environmental quality, to establish technical cells for environmental assessment.

Eco-development

Most of the programmes for environmental management deal with pre-planning for eliminating or at least minimizing environmental degradation. One of the major objectives of eco-development was restoration of already degraded ecosystem through practical field scheme such as land reclamation, afforestation and cleaning of water bodies, etc.

The Eighth Five-Year Plan

The scenario of environment and forests continues to cause concern and the destruction and degradation of forests are taking a heavy toll of our soil and water resources. In the Eighth Plan, in addition to a number of national level bodies which have been constituted by the Ministry on Environment and Forests, the Planning Commission had set up several expert groups/committees to formulate long-term sectoral policies and to reconcile the conceptual confrontation between environment and development. Submission of an environmental statement by the polluting units to the concerned State Pollution

Control Board was made mandatory through notification under Environmental Pollution Act (1986).

Adoption of Clean Technologies in Small Scale Industries

This scheme seeks to promote the development and adoption of clean technologies, including waste reuse and recycling and to link research and development to prevent pollution.

Environmental Statistics and Mapping

Under the scheme, activities relating to collection and analysis of environmental data and its direction on an atlas were thought to be carried out.

Some other important programmes were:

- World Bank Assisted Industrial Pollution Control Project (Phase I and II)
- Strengthening EIA
- National River Conservation Programme
- Integrated Afforestation and Eco-Development Project

The Ninth Five-Year Plan

One of the objectives of the Ninth Five-Year Plan was to ensure environmental sustainability of the developmental process through social mobilization and participation of people at all levels.

Forests

For ecological sustainability, different afforestation programmes and all round development of the villages for the well-being of forests were planned to be initiated. Involvement of NGOs in areas of awareness building and community education was planned to be encouraged.

Issue-Specific Programmes

- People's involvement and role of information
- Citizen's Monitoring Committees were established under the National River Conservation Programmes
- People at large and the university systems, particularly science, engineering and medical facilities were planned to be involved in monitoring and enforcement work

Area-Specific Programmes

- National River Conservation Programme (launched in 1995 to clean stretches of major rivers in the country)
- National Lake Conservation Programme (the plan was taken up since 1997 to clean and restore major lakes of the country)
- Integrated Development Programmes for Himalayan Regions
- Islands Conservation Programmes

Sector-Specific Programmes

- Strengthening Central Pollution Control Board (CPCB)
- Industrial Pollution Control and Prevention Project

The Tenth Five-Year Plan

The Tenth Plan is mainly a coordination of some previous projects with some new initiatives. The plan places emphasis on different pollution problems continuing from the Ninth Plan as follows:

Air Pollution

- The CPCB has established a national air-monitoring network covering 290 stations spread over 90 cities and towns. To reduce vehicular pollution, emission standards and restrictions were imposed for petrol and diesel-driven vehicles.
- Guidelines have been made for setting up of industries.
- Twenty-four critically polluted areas have been identified across the country and an action plan has been devised for the control of pollution in these areas.

Water Pollution

- Emphasize was given on mainly ongoing projects, viz., National River Conservation Plan and National Lake Conservation Plan. No other new initiatives were made in this plan.

Solid Waste

- Master plan for municipal solid waste.
- The Bio-medical Waste Management and Handling Rules have been emphasized in new form.

- While emphasis in the Tenth Plan was on completing the ongoing process, there are some new initiatives in the plan such as:
- The scope of Common Effluent Treatment Plants scheme was enhanced to cover assistance for modernization and capacity expansion of existing plants.
- Scheme with international cooperation (for watershed management, water and energy sectors, bio-diversity) with the financial and technical help from India-Canada Environmental Facility, Global Environment Facility, etc.
- Schemes under the Clean Development Mechanism: Direct measurement of temperature as well as visible effects like, rise in sea levels.

Table 1

Sector-wise Outlays in India's Planning

Head of Development Sectors	*Total Outlays (Rs. in crores)*				
	6th Plan	*7th Plan*	*8th Plan*	*9th Plan*	*10th Plan***
Agriculture and Allied Activities	5695	10573	22460	42462	58933
Special Area Programme	1480	16978	12052	55420	20879
Irrigation and Flood Control	12160	54821	51850	74686	10335
Rural Development	5363	3145	3827	3649	121928
Energy	26535	22461	25100	222375	403927
Industry and Minerals	15017	15245	35025	65748	58939
Transport	2412	22971	32561	119373	225977
Communication, Information and Broadcasting	3135	6472	7002	47280	98968
Science and Technology and Environment	–	2466*	5635	–	30424
Environment and Ecology	–	427	–	18459	–
General Economic Service	–	–	–	14580	38630
Social Service	14035	29350	30547	183273	34739
Others	801	1686	2015	12496	16328
Total	97500	180000	569085	859200	1525639

* Excluding environment
** Projected figures by Planning Commission, Government of India

Source: Planning Commission Report: 6th to 10th Five-Year Plans, Government of India.

- Six projects mainly of non-conventional energy sources having been selected for funding from Netherlands.
- The state of environmental scheme: The central aim of the scheme is assessment of physical parameters, sensitization of developmental planners and introduction of corrective measures in mainstreaming environmental concerns.

Important Laws Enacted for Environmental Protection

A well-developed framework for environmental legislations did not exist in India until after the 1972 UN Conference on Human Environment (Stockholm) took place (TERI, 2003). After that India entered a new era in the field of environmental legislations by enacting laws specifically aimed to deal comprehensively with environmental issues.

Some of the important legislations enacted in this connection so far include:

- Wildlife Protection Act, 1974
- Water (Prevention and Control of Pollution) Act, 1974
- Water Cess Act, 1977
- Forest Conservation Act, 1980
- Air Pollution Control Act, 1986
- Certain major Acts, such as Motor Vehicles Act, 1938 were amended in 1988, to address the concern of environment
- Forests Act of 1980 and National Forest Policy of 1988

The administration of these laws had been entrusted to agencies created under the respective statutes. Some of the Acts were also reviewed and amended according to the recommendations of Tiwari Committee (Pal, 1999).

Some New Laws Enacted

After the amendment and reviewing of the new laws and keeping in view the present environmental problems there are some new laws enacted such as:

- The National Environment Appellate Authority Ordinance, 1997
- Bio-medical Waste (Management and Handling) Rules, 1998
- The Policy Statement of Abatement of Pollution, adopted in 1992 (Batta and Bhatti, 2001)

- Efforts to save biodiversity around Gulf of Manner, 2002
- Supreme Court's nod for Ganga Action Plan, 2002
- Modified Water Policy adopted in 2002
- Government launches Special Water Management Plan, 2002
- KWIP (Kolkata Water Improvement Project), finalized in 2002

Criticism of the Five-Year Plans

Planning in a rational way constitutes an essential tool for all-round development as well as to protect the environment. We can say that a development process must be ecologically viable. Yet on the policy level, the Government of India has been focusing many environmental measures. But what lacks in the planning is a holistic approach and coordination between planning, law and implementation. After having an overall idea about the planning of India we can see that government had taken several steps for overcoming the environmental problems in different five-year plans. Though in first three five-years plans there were less emphasis on environmental concern. The Sixth Plan first drew attention on environmental planning. But less effort was given to financial help to the plan. Coming to the Seventh Plan it is evidenced that for the first time a provision for separate sector and financial outlay was made in the plan. Again, in the Ninth Plan, provision for separate environment sector was evidenced. But, in the latest plan, i.e., the Tenth Plan, we could not see any separate provision for environment, and it is merged with science and technology sector (see Table 1). That is the problem with India's planning. They are not maintaining a continuous plan for the environment. It is very much needed to establish a separate sector for environment and ecology, and to inculcate the financial resource also. Fact is that environmental policy in India has relied only on command and control instruments.

Throughout the world, the feelings of desperation and failure are prevalent in the environment policy area, partly because of an ever-increasing gap between policy goals and policy performance (Singh and Sharma, 2004). Yet, there are sufficient environmental regulations to cover the issue to a reasonable extent. The question is certainly not about numbers but of designs. It is not the absence of policy, but the inadequacy of the design. Because faulty policy designs lead to administrative problems, which in turn lead to policy failure.

On the other hand, we have to recognize the fact that problems generated by environmental crisis in our cities and villages cannot be solved by legislative measures, devised solely for environmental protection. But they must be dealt with by means of environmental planning for maintaining ecological and biological balance, essential for the survival of the human being. In our national developmental planning, it is essential to adopt an integrated and coordinated approach to ensure, the compatibility of development with need for protection and improvement of human environment for the benefit of our people.

References

Batta, R.N. and Bhatti, J.P (2001): "Environment Policy in Developing Countries", *Man and Development*, 23(4): 7-21.

Fernandes, B.G. (1996): *Planning in India: Challenges and Opportunities,* Anmol Publications Pvt. Ltd., New Delhi.

Ministry of Information and Broadcasting (2005): *India 2005: A Reference Annual*, Ministry of Information and Broadcasting, Government of India.

Pal, Chandra (1999): *Environmental Pollution and Development: Environmental Pollution Policy and Judiciary*, Mittal Publication, New Delhi.

Planning Commission, Govt. of India: *Ten Five-Year Plans Reports (First to Tenth Plan), from 1951 to 2002,* Planning Commission, New Delhi.

Singh, P.P. and Sandhir Sharma (2004): *Environment and Pollution Education*, Deep and Deep Publications, New Delhi.

TERI (2003): *Teri Energy Data Directory and Year Book (TEDDY)*, Tata Energy Research Institute, New Delhi.

II Food, Water and Sanitation

5 Population vs. Food: Who is Winning the Race?

Alka Parikh

Journey towards self-sufficiency in food was not an easy one for India. We traversed from the periods of famines during the British era, to importing PL 480 wheat for years, to dramatic increase in wheat production in the 1970s, to the spread of agricultural technology in semi arid tropics and crops other than wheat in 1980s. There have been wide fluctuations in agriculture production, but by and large, India has become self-sufficient in food.

However, there is no surety that gains in productivity would be sustained. First concerns regarding sustaining the increased food production were expressed during 1980s. The concern was based on environmental effects of the High Yield Variety (HYV) technology. Two major points of concern were the following: (1) Use of chemical pesticides and fertilizers can affect the bacteria in the soil thereby reducing the capacity of the soil to regenerate its nutrients. The soil thus would become less and less fertile with time. (2) With excessive irrigation and inadequate arrangements for drainage, problems of waterlogging would occur, increasing the soil salinity, making the soil barren. Thus, yield per hectare was supposed to decline with time. It was feared that the increased production would not be sustained in the long run. The yields have not really declined when we look at India as a whole, but the growth rate in yields might have gone down due to environmental reasons.

The 1990s brought in its own set of problems. With structural adjustment initiatives implemented, the focus shifted to fiscal measures and their impacts on developmental process. Roy and Pal (2002) show that public investment in agriculture has declined considerably. They caution that such decline in investment might affect agriculture adversely. Unlike the environmental factors, the effect could be felt not in the long run, but in near future itself.

By the time the 1990s ended, many authors reported that agricultural growth rates in India are slowing down (Rao and Jeremy, 2000; Dev, 1998; Mallik, 1997; Sawant and Achuthan, 1995; Sawant et al., 2002). If the rate of growth of food grain production slows down, and that of the population hardly declines, have we really won the race, as far as food availability is concerned? How justified are the fears that soon we might be losing the race again?

These are the questions we try to address in this paper. One problem with the studies mentioned earlier is that they compare the growth experience of the 1980s with the 1990s. We maintain that the notion of measuring agricultural growth by dividing the performance into decades is erroneous (Parikh et al., 2005). Agriculture production in India fluctuates widely from year to year (Figure 1). It is not correct to artificially put a cut-off point (like 1981 or 1991) in a widely fluctuating time series and then try to discern the trend from that point onwards. It would erroneously give an increasing or decreasing trend, whereas the time series over a few more years might show a completely different trend. We feel that when we have the advantage of having the time series data for at least 30 years, it would be better to look at the entire time series and find whether cutting it off at 1990 or 1980 makes any statistical sense.

Trends in Agricultural Growth

As there are wide fluctuations in the production, moving average for three years was used to smoothen the food grain production data for India for the years 1972-73 to 2000-01. The resulting curve is given in Figure 1. The curve shows rising trend, and the visual inspection does not seem to give any indication of there being a kink in the curve around the 1990s.

To probe it further, we looked at the rates of growth experienced in the food grain production. When the annual rates of growth are graphed for the years under consideration, we find that they look smaller in 1990s, but there is also much less fluctuations in the 1990s, (Figure 2), hence the mean for the series might not have changed much in the 1990s. However, that also does not give a precise picture. A regression analysis is the best way to determine the rates of growth.

Figure 1

Food Grain Production in India

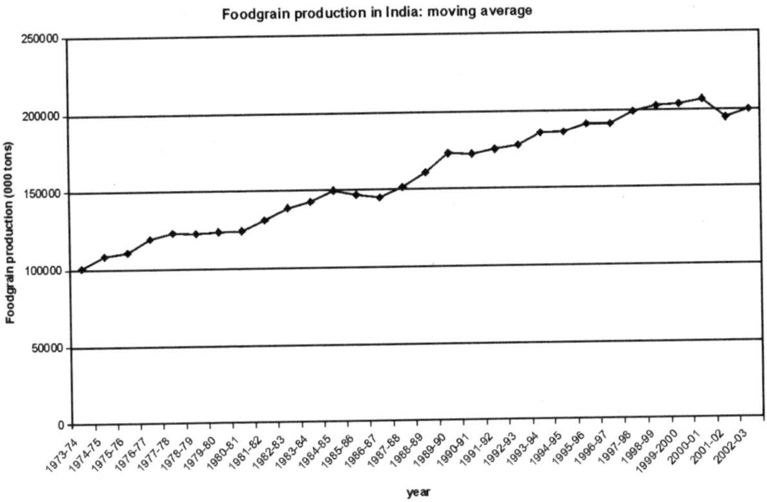

Source: CMIE, "Agriculture", 2002.

Figure 2

Rate of Growth of Food Grain Production in India

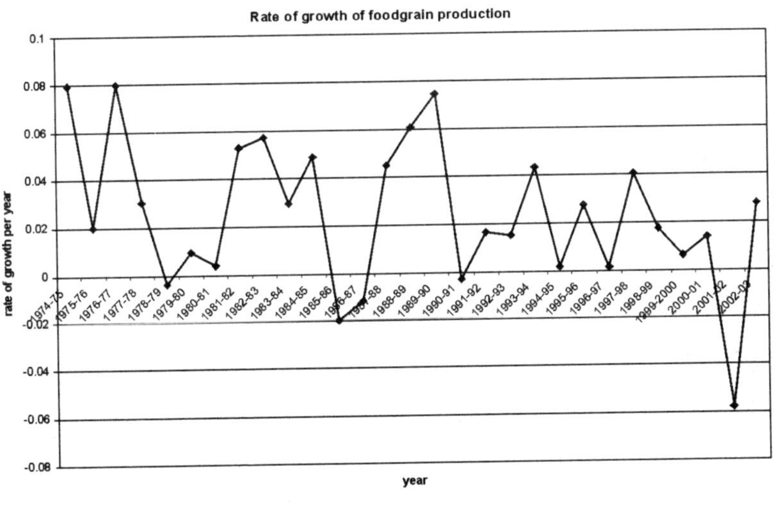

Source: CMIE, "Agriculture", 2002.

Our regression analysis was carried out in two phases: First we ran a semi-log regression with ln (output) as the dependent variable and time as the independent variable to find out the rate of growth over last 30 years. The compound rate of growth of was found to be (2.5% p.a.) for food grains production in India from 1972-73 to 2001-02. The results of this regression are given in Table 1 (a).

To determine whether the rate of growth has indeed slowed down in 1990s, we respecified the model (Hatekar and Dongre, 2005; Gujarati, 1988). Dummies were introduced in the regression to represent the decades and together with time trend, interactive term of the dummies and time were also included. The model, thus was specified as follows:

Log (production) = intercept + α_1 (dummy for 1980s) + α_2 (dummy for 1990s) + β (time trend) + θ_1 (dummy for 1980s * time trend) + θ_2 (dummy for 1990s * time trend) + error term.

The results of the regression analysis are given in Table 1 (b).

Table 1

Results of Regression Analysis

a. Results of semi-log regression equation for food grain production	
R-square	0.980878
Adjusted R-square	0.980143
Intercept coefficient	-39.3413 (t statistics: -28.0207)
Time trend coefficient	0.025811(t statistics: 36.52007)
b. Results of semi-log regression equation with dummies for each decade	
R-square	0.987491
Intercept coefficient	-46.8156 (t statistics: -5.77902)
Dummy 1980s	3.24087 (t statistics: 0.325162)
Dummy 1990s	21.90354 (t statistics: 2.193864)
Time trend coefficient	0.029588 (t statistics: 7.21888)
D1980s*time	-0.00164 (t statistics: -0.3252)
D1990s*time	-0.01101 (t statistics: -2.18663)

The results show that the pace of growth in agriculture has indeed slowed down in the 1990s. The rate of growth has almost dropped down to 1.9 per cent, per annum (it dropped further in 2000s). This is indeed a matter of concern.

The diet of the poor in India still consists mainly of cereals. Hence, trends in cereals production are crucial for determining food security of Indian population. In addition to the total food grain production, it is thus important to look at the cereals. It is also important to consider the cereals production since Sawant (2002) claims that the growth rates of cereals are affected the most.

We used the same model specification as used for food grain production for evaluating trends in wheat production in Punjab, rice production in Tamil Nadu and Andhra Pradesh and jowar production in Maharashtra. The only change made was to make the yields as an indicator of growth rather than total production of that cereal.

Our results (please see the Appendix) showed that the wheat yields in Punjab grew at the rate of (2.6% p.a.) between 1971-72 and 1998-99. There does not seem to be much statistically significant difference between the growth performance in 1970s and 1980s, maybe because the technology was still spreading to the fields of small and marginal farmers. However, in the 1990s, the growth rates have slowed down (the time and 1990s dummy was statistically significant); they have fallen to (1.3% p.a.). This might have happened because with the given technology, we might be reaching the maximum levels that are possible to produce, leaving little scope for unlimited increases in productivity. Improper irrigation management is also blamed for the declining growth rates: problems of waterlogging are largely related with lack of proper drainage and tendency to over irrigate. It would be proper to state that now in Punjab, the main concern is to sustain the high yields, rather than trying to better them. In other words, gains in production cannot be expected to come much from Punjab and Haryana anymore, unless the production technology changes.

Two major rice producing states, Tamil Nadu and Andhra Pradesh, were evaluated for their performance in yield growth. There is a distinct kink in the yield curve of Tamil Nadu: the rice yields hovered around 1,800-2,000 tonnes per hectare from 1971-72 till 1984-85, but they started rising steadily after that. By 1989-90, the yields had increased to about 3,000 tonnes per hectare. Since then, the yields have been around 3,000-3,500 tonnes. There is no distinct trend after 1990. As opposed to Tamil Nadu, yields have been constantly growing in Andhra Pradesh, although the yields per hectare in Andhra are much lesser than those of Tamil Nadu (2,780 kg/hectare in 1998-99). The yield growth rate has been about (2.5% p.a.) since

1971-72 to 1998-99. But the yield growth rate has slowed down in Andhra also to barely 0.08 per cent in the 1990s. Given the production possibilities with the existing technology, there is scope to increase the yields of Andhra Pradesh at least to the levels of Tamil Nadu. There should be measures to reverse the trend of decelerating growth.

We looked at one coarse cereal also: jowar in Maharashtra. Acreage under jowar has been steadily declining; cash crops have been taking over the fields from jowar. However, its yields have increased over years (the growth rate in yield has been 1.8 per cent per annum from 1974-75 to 1998-99), the growth was slower in 1980s compared to 1970s; it seems to have recovered a little in 1990s. Coarse cereals might hold out some hope for increasing national food grain production.

The yield growth rates in production of cereals have been stagnating or declining by the time 1990s ended. The areas where yields have been traditionally low are the semi-arid tropics, where production gains historically have been limited due to water shortage. Since the full potential of the technology is not yet attained, there is a hope that we can still increase food grain production there. Watershed management projects, started by both the governments and the funding agencies, hold out some hope for these areas.

Population and Food Grain Production

Before we go into the implications of the above mentioned results on food security, we would like to emphasize that this paper concentrates only on food availability issues. The distributional aspects are not dealt with here.

Table 2

Growth Rates in Population and Food Grain Production

	Population (in million)	*Index*	*Food Grain Production*		*Per Capita Food Grain Production*
			(million tonnes)	*Index*	
1950-51	363	100	51	100	140
1960-61	442	122	82	161	185
1970-71	551	152	108	213	196
1980-81	689	189	130	255	188
1990-91	852	234	176	347	206
2000-01	1033	284	196	386	189

Source: indiabudget.nic.in/es2001-02/chapt2002/chap515

Table 2 gives the comparison between population growth rates and growth rates in food grain production. In spite of considerable growth in food grain production in last few decades, the per capita food availability has not changed much, maybe due to wide variations in food grain production. The per capita food availability has increased from 140 kg per annum in 1950-51 to around 185-200 kg. However, this range has remained more or less the same from 1960s till 2000-01. Half kg of food grains per capita per day is sufficient for a person. Hence, at present, the problem of food availability does not exist. The question is: what could the future hold in store for us?

We considered two scenarios for the future: (a) sustaining food grain growth rate; and (b) declining food grain growth rates. The population growth was assumed to remain around 1.9-1.8 per cent p.a.

Sustaining Food Grain Growth Rates

The results given in Section 2 show that although Indian food grain production is growing at the rate of 2.5 per cent per annum over last three decades, the growth rates are slowing down now. There is gradual slowing down of rates of growth in wheat production in Punjab or in rice production in Andhra Pradesh. We feel that in the absence of major technological breakthrough in agriculture in last two decades, this was bound to happen. The HYV technology led to phenomenal increases in yields for a decade and a half. Now, the HYV technology is well spread and well used. Unless some new technological development happens, the yields might stagnate since the upper limit of the production possibility frontier that was possible from the HYV technology might be reached.

The heartening fact is that the yield growth rate has slowed down but it has not stagnated yet, as far as total food grain production for the country goes. In other words, the food grain production is still increasing. Also, if the watershed management projects are successful in supplying water to the semi-arid tropics and if infrastructure development spreads to interior regions, some gains in productivity in these regions can still happen. At the same time, success in BT cotton is reported to have increased cotton production. There might be similar technological breakthroughs in store for cereals also in the near future.

The per capita food availability was just enough in 1990s. The race between the food grain production and population growth was not yet lost. With the positive policies for agriculture by the Manmohan

Singh's government, leading to further infrastructure and technological development might actually help in sustaining the growth rates in food grain production. We might not be able to increase the growth rates back to 2.5 to 3 per cent p.a., but even if we sustain the growth rates of 1.8 to 1.9 per cent, we would still not be falling behind the population growth rates. We would be able to remain self-sufficient in food for at least one more decade.

Declining Food Grain Growth

It took Punjab 15 years of green revolution to show the first signs of slowing down. Similarly, the slowing down has been gradual for all cereals. If the food grain production growth rates were to decline, it is not unrealistic to project that the decline would be gradual. Given the fact that India is considered as one of the fastest growing nations, the increased incomes of Indians (and especially the increased foreign exchange) would come handy in importing food grains from abroad. And, as the decline would be gradual, only small quantities would need to be imported. India might actually be able to afford such imports.

However, if the yield growth rates do not decline just gradually, but start stagnating soon, India will have to import considerable quantities of food grains. It would be economically crippling and politically disastrous (arm twisting by others becomes much easier when the nation is heavily dependent on others for food). Many poverty alleviation programmes are based on food grain distribution at present, e.g., PDS, mid-day meals, food for work programme, drought relief programme, disaster management programmes, etc. It was possible to introduce them because we have become self-sufficient in food production. The country would not be able to afford such programmes, if food grains are imported from outside. Thus, future might be very bleak for the poor and deprived.

Knowing that there is distinct possibility of this becoming true, steps need to be taken seriously and quickly to put agriculture back on growth path. Investments in agriculture should increase, allotments for research and development should be prioritized, infrastructural facilities (including credit, roads, communication, markets, and such) should be developed and government should assume a serious supportive role for agriculture. A fast growing nation should support its basic primary sector as much as possible.

Addendum

Demand for food does not come merely from physiological factors (although they are the most important ones), but from other factors also, including growth in incomes. Per capita incomes have been rising in India. That could lead to increase in demand for food. Moreover, the tastes of the population have been changing. Consumption of meat is becoming more common among youngsters of the communities that have been traditionally strictly vegetarian. Such change in tastes and preferences can lead to increase in demand for meat and hence increase in demand for animal feed. Animal based diet requires about six times as much resources as the vegetarian diet. Such increases in demand for cereals are not taken into account while making these calculations. This paper considers only physiological factors while comparing the growth rates of food grains and population.

Conclusion

Fears are expressed that agricultural growth rates are slowing down. Even with better model specification, we found that the rates of growth of food grain production have indeed slowed down. Signs of slow down are distinctly clear for wheat yields in Punjab. Andhra rice yields are also slowing down, in spite of the fact that they have not even reached the maximum possible yields. Yields of rice in Tamil Nadu are increasing with fits and starts; it is difficult to say whether they would increase or stagnate in the near future. In Maharashtra, the trend has been fluctuating, and the acreage under jowar has also been declining over time.

We argue that if the growth rates have finally fallen down to more realistic levels now and if we could sustain them, we would still remain self sufficient in food. If the growth rates actually slow down considerably, we would have a serious problem at hand. In the face of such eventuality, the sooner we take corrective measures, the better.

Appendix

Results of semi-log regression equation for Andhra Rice	
R-square	0.982203
Intercept coefficient	-45.2197 (t statistics: -5.1605)
Dummy 1980s	13.7225 (t statistics: 1.1045)
Dummy 1990s	36.3400 (t statistics: 2.0731)
Time trend coefficient	0.0266 (t statistics: 6.0051)
D1980s*time	-0.0069 (t statistics: -1.10025)
D1990s*time	-0.0182 (t statistics: -2.0717)

Results of semi-log regression equation for Maharashtra Jowar	
R-square	0.976013
Intercept coefficient	-327.134 (t statistics: -16.6032)
Dummy 1980s	370.9691 (t statistics: 16.2507)
Dummy 1990s	317.2778 (t statistics: 13.2579)
Time trend coefficient	0.16886 (t statistics: 16.9108)
D1980s*time	-0.18767 (t statistics: -16.2462)
D1990s*time	-0.1605 (t statistics: -13.2776)

Results of semi-log regression equation for Tamil Nadu Rice	
R-square	0.97036
Intercept coefficient	-2.4072 (t statistics: -0.278)
Dummy 1980s	-108.674 (t statistics: -8.8519)
Dummy 1990s	34.4116 (t statistics: 1.19868)
Time trend coefficient	0.00505 (t statistics: 1.1530)
D1980s*time	0.0548 (t statistics: 8.8446)
D1990s*time	-0.0171 (t statistics: -1.9627)

Results of semi-log regression equation for Punjab Wheat	
R-square	0.993579
Intercept coefficient	-43.2579 (t statistics: -10.8476)
Dummy 1980s	-2.8448 (t statistics: -0.5032)
Dummy 1990s	24.5033 (t statistics: 3.0718)
Time trend coefficient	0.0258 (t statistics: 12.8074)
D1980s*time	0.00145 (t statistics: 0.5076)
D1990s*time	-0.0123 (t statistics: -3.0656)

References

Dev, Mahendra (1998): "Regional Variations in Agricultural Performance in the Last Two Decades", *Indian Journal of Agricultural Economics,* Vol. 53, No. 1, January-June.

Gujarati, D.N. (1988): *Basic Econometrics,* McGraw-Hill International Editions, Economics Series.

Hatekar and Dongre (2005): "Structural Breaks in India's Growth: Revisiting the Debate with Longer Perspective", *Economic and Political Weekly.*

Mallik, Jayant Kumar (1997): "Growth of Agriculture in Independent India:50 Years and After", *RBI Occasional Papers,* Vol. 18, Nos 2 and 3 special issue, June and September in Economics, Department of Economics, University of Mumbai.

Parikh, Alka, Sarthi Acharya and Maithreyi Krishnaraj (2005), "Impact of Maharashtra's Agricultural Policies on Women Farmers: A Gender Budgeting Analysis", *Discussion Paper Series 8,* HDRC, UNDP, India, April, 2005.

Rao, V.M. and P.D. Jeromi (2000): "Modernising Indian Agriculture: Priority Tasks and Critical Policies", *Development Research Group, Study No 21,* Department of Economic Analysis and Policy, Reserve Bank of India, August 29.

Roy, B.C. and S. Pal (2002): "Investment, Agricultural Productivity and Rural Poverty in India: A State Level Analysis", *Indian Journal of Agricultural Economics,* Vol. 57, No. 4, October-December, 2002.

Sawant, S.D. and C.V. Achuthan (1995): "Agricultural Growth Across Crops and Regions: Emerging Trends and Patterns", *Economic and Political Weekly,* March 25.

Sawant, S.D., Vaidehi Daptardar, and Sandhya Mhatre (2002): "Capital Formation and Growth in Agriculture: Neglected Aspects and Dimensions", *Economic and Political Weekly,* March 16.

Sawant, S.D. (2002): "Indian Agriculture: Past Developments and Policies for the Future", *Prof. M.L. Dantwala Monograph Series,* Department of Economics, University of Mumbai, October, 2002.

6 Water Resources in India: Issues, Concerns and the New Environment Policy

Sumati Kulkarni

Water has always been one of the crucial factors shaping the course of human civilization. All the ancient civilizations flourished on the banks of big rivers. Water is essential for the survival and sustenance of life of human beings, animals, birds, fish as well as all flora and fauna. Water being the source of irrigation as well as of electricity, has influenced the pace of development of nations during the agricultural and industrial era. In recent decades, however, with explosive population growth, rising aspirations, unplanned urban growth and increasing environmental stress, water is considered to be a matter of serious concern. It has been predicted that the next world war will be fought not on the oil issue but on the issue of access to water.

In India, in spite of the six decades of developmental planning, agriculture is still a gamble in monsoons. Every year the country has to face severe droughts in some parts and terrible floods in some other parts. There is growing realization that it is not a matter of just water shortage due to population growth. Water management has many dimensions and it is now a very complex issue, mainly because of various problems such as inter-state water disputes, salinization of land due to large irrigation projects, large scale water contamination due to huge industrial and human waste, etc. It is further complicated by the contrasting stands taken by various lobbies, the groups involved and the vested interests.

The paper proposes to assess the magnitude of the problem and to discuss the demographic as well as non-demographic dimensions of the water problem to bring out the complexity of the issue.

Objectives

- To examine the demand and supply aspects to assess the magnitude of the problem.
- To discuss various dimensions of the problem regarding the quantity and quality of water resources in India and their impact on environment.
- To review briefly the strategies adopted/recommended to deal with these problems.
- To discuss various perspectives, controversies and dilemmas regarding water management.
- To discuss the approach adopted in the national water policy and to examine how far the new environmental policy addresses the complexity of the whole issue.

Magnitude of the Problem

1. Annual per capita availability of surface water in India has declined from 5,410 cubic metres in 1951 to 1,902 cubic metres in 2001 and is estimated to decline to 1,235-1,451 cubic metres by 2050.
2. Annual per capita availability of utilizable surface water has declined from 1,911 cubic metre in 1951 to 672 cubic metre in 2001 and is estimated to decline to 421 cubic metre by the year 2050 (Kumar et. al., 2005).
3. The norm stipulated in the Tenth Plan: (1) 40 litres per capita per day of safe drinking water within a walking distance of 1.6 km or elevation difference of 100 metres in hilly areas, and (2) at least one handpump/spot source per 250 persons.
4. Per capita availability varies from 10 kilolitres to 50 kilolitres per annum for different seasons and regions in India.
5. 50 per cent of irrigated agriculture is based on groundwater, 85 per cent of rural drinking water comes from groundwater (NCIWRDP Report, 1999).
6. 81 per cent of rural and 19 per cent of urban households have no toilet facility (NFHS-2).
7. One among every eight households in India does not have water source within 1 km. Walking distance (i.e., about 30 minutes for round trip to fetch water). Among ST households or among households with low standard of living, one in every five households is in this category (NFHS-2).

Floods and Droughts

1. One-third of population is affected by floods or drought every year.
2. Floods affect eight major river valleys, spread over 40 million hectares and affect 260 million people resulting into loss to property, crops, people worth Rs 2,400 crores on an average per year. In addition, government incurs expenditure on relief measures worth Rs 1,200 crores.
3. Droughts affect 86 million people spread over 116 districts in 14 states.

Environmental Impact and Water Quality

1. Nearly 70 per cent of the available water in India is polluted.
2. 73 million work days are lost due to water related diseases.
3. Per day, about 873 million litres (MLD) of waste water was drained into Ganga before the Ganga Action Plan was initiated.
4. Overtoxication of water for irrigation and urban consumption leaves several rivers dry for most of their course denying dilution of pollutants flowing into the river. Yamuna has no water downstream of Tajewale.
5. Indiscriminate, uncontrolled discharge of industrial effluents and high concentration of oil and toxic substance make Damodar river water unfit for drinking and even agricultural use.
6. The extent of waterlogged land and salt affected land is 2.46 million hectares and 3.30 million hectares respectively (Working Group of the Ministry of Water Resources, 1991).
7. Sankosh project proposed by the West Bengal government, having a canal system 180 km long and 120 metres wide threatened the rich forests of the Buxa tiger reserve.
8. Slow death of Kashmir's Dal lake is threatening the livelihood of 50,000 fisher folk.
9. India's groundwater resources are about ten times its annual rainfall but with 170,000 tube wells added every year during 1970s, the water table has been declining in many areas.

Supply and Demand Situation: Available Estimates

Available and Usable Water Resources in India

According to the estimates of the working group appointed by the National Commission for, Integrated Water Resources Development

Plan (NCIWRDP, 1999) India gets on an average 4,000 billion cubic metres (BCM) of water per annum by way of rain and snow but, the annual water resources measured in terms of flows at terminal points of the river system are only 1,953 BCM. In addition, 432 BCM groundwater is available. Out of this, only 1,086 BCM water is usable (690 BCM of surface water and 396 BCM of groundwater). However, the present use is about 600 BCM.

Therefore, nearly 486 BCM of water is unutilized and can be harnessed.

Figure 1

Demand Situation

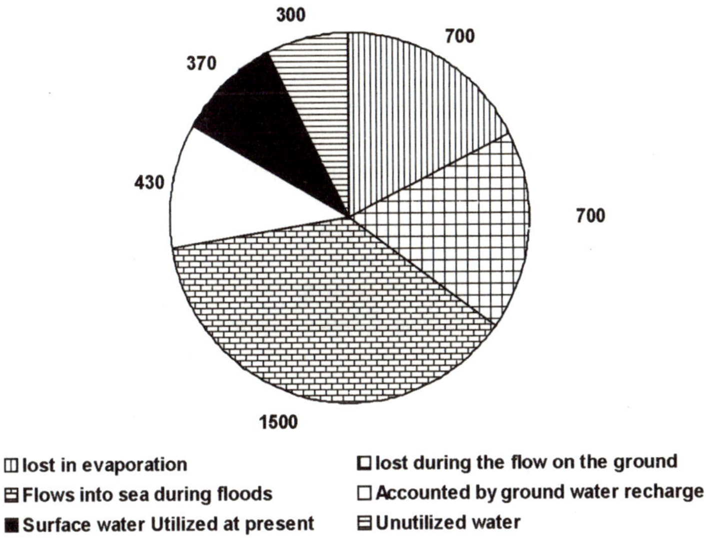

Demand for water comes from the domestic sector (for drinking, washing, bathing and sanitation), from the agricultural sector (for irrigation), from the industrial sector including construction, from the power sector, from the transport sector (for navigation) and for some miscellaneous purposes. In addition, it is also essential to take into account some quantity, which is usually lost due to evaporation. The future water needs of any country depend on its population size and its level of urbanization because usually water consumption in urban areas is likely to be higher due to differences in the lifestyle.

For various types of uses NCIWRDP has estimated India's future water requirements for the years 2010, 2025 and 2050 under the following alternative assumptions:

Assumption 1: L. Low variant of the U.N. Population projections, 1994 and low rate of urbanization.

H. Population as per the projections by Visaria and Visaria and high rate of urbanization.

Assumption 2: Domestic use requirements based on the following norms:

	Year		
	2010	*2025*	*2050*
	(litres per capita per day)		
Class I cities	220	220	220

Requirements of water for various other sectors were projected by assuming the pattern consistent with the present composition of sector-wise requirements and by making appropriate assumptions regarding the level of requirements of other sectors (details available in the report of the working group of NCIWRDP). Figure 2 shows the projected

Figure 2

NCIWRDP Estimates of Future Requirements (BCM)

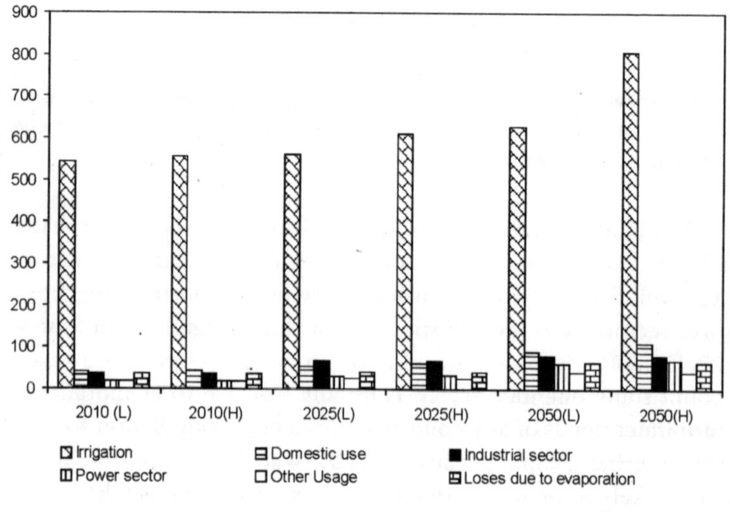

total water requirements and their break-up by sector-wise total requirements. It shows that irrigation accounts for major portion of the total requirements followed by the requirements of the domestic and industrial sectors which form a much smaller component than the irrigation requirements. It is the result of the present pattern as well as the due emphasis placed on the consideration for food security in making the assumption for future.

Table 1 gives the total water requirements for all uses for the years 2010, 2025 and 2050 as projected by NCIWRDP. Two other estimates for years 2020 and 2025 are also given. Though they are not strictly comparable with the estimates given in the table due to the difference in the assumptions, they give broad idea about the range.

Table 1

Total Future Water Requirements for All Uses (BCM)

Year	*L*	*H*
2010	694	710
2025	784	850
2050	973	1180

The demand projections by the NCIWRDP were found to be on the lower side by some, while some found them to be sufficiently high to indicate the crisis, as shown by the following viewpoints.

No doubt the gap between the demand (even the low variant) and supply does underscore the need to take immediately all possible steps to increase the supply and control the demand. In addition, measures to control water pollution and ensure the water quality are needed so that all the available supply can be utilized. Various short-term and long-term remedies adopted/suggested in this context are summarized below.

Short-Term and Long-Term Remedies Suggested/Adopted

Supply Side Solutions

1. Construction of dams (major and minor irrigation projects).
2. Maintenance of existing irrigation projects to utilize the full potential of the existing irrigation projects.

3. Linking of rivers.
4. Rainwater harvesting.
5. Flood control, storing floodwater for use in non-monsoon months/for distributing it to draught prone areas.
6. Clearing the inlets and outlets of all water bodies.
7. Groundwater management.
8. Water recycling.
9. Drainage management.
10. Afforestation.

Demand Side Solutions and Measures for Water Quality Improvement

1. Population control. A comprehensive long-term measure.
2. Lifestyle changes. In India, some religious practices like mass bathing in rivers as in Kumbhmela and disposal of dead bodies by throwing them in the river have made river water unsuitable for use at many places.

   Though individually each of these measures appear to improve the water situation, it is essential to examine (a) whether they have the capacity to achieve the desired results in terms of demand, supply or quality and at what cost, (b) whether due to them, better off sections are likely to benefit at the cost of weaker sections, and lastly (c) whether because of them the present generation is likely to benefit at the cost of future generations. In short, there is a need for an integrated approach, incorporating the above strategies on the following guiding principles of efficiency, equity and sustainability. Some of the solutions which were acceptable in the past are now being questioned on the grounds of these principles. Water management issue faces today a big dilemma because of the controversies arising out of different perspectives in this respect.

Perspectives, Controversies and Dilemmas

Large Dam's Controversy

In India, right from the First Five-Year Plan, large-scale irrigation projects like Bhakra Nangal, Hirakud and Damodar valley projects have been the main strategy for agricultural development and electricity generation for industrial development. Large dams were described as shrines of modern India by our first Prime Minister Pandit Jawaharlal Nehru. Half of the existing large dams in India were constructed during the seventies and eighties. Two-thirds of the installed hydro-power capacity in India by the end of last century is attributed to large dams. They have played an important role in the growth of food production from 51 million tonnes in 1951 to nearly 200 million tonnes at the end of the century. Yet today the prevalent view is in favour of small and medium irrigation projects. Increasing criticism of large dams, especially since Narmada movement, is not only due to the long time and huge cost involved but also due to the large scale displacement of people and adverse impact on environment. Failure to get the projected benefits, poor pricing of irrigation water resulting into the inability to generate revenue for the proper maintenance of dams are the other issues involved. It is observed that mostly rich farmers have reaped the benefits of these dams, leading to vested interests and the whole politics of water in rural areas.

Linking of Rivers Controversy

In India, there is huge wastage of water annually when big rivers in the North are flooded during monsoons (or for Himalayan region rivers during summer, when the snow melts). The Grand Canal Scheme linking Ganga or Brahmaputra to Godavari in the South, to distribute the surplus water in the North to the draught prone areas in the South has been mooted by various experts like Rao, Dastur from time to time. Recently, it is being discussed again due to the viewpoint expressed by Supreme Court in its favour. Its desirability and viability, however, has been questioned by experts on techno-economic grounds as well as on environmental grounds. There is an apprehension that such a grandiose project will not only involve huge cost of crores of rupees but will lead to large scale displacement of people, major environmental disturbance, drastic changes in the morphology and quality of water and severe climatological changes. All this can have tremendous adverse impact on wildlife, aquaculture, flora and fauna. Besides, with all this there is no guarantee that it will improve the conditions of those areas which suffer from water scarcity. Lastly people's feelings on inter-state water disputes like Cauvery issue between Tamil Nadu and Karnataka are so intense that such a project is almost impractical. In case of Ganga, it will have international implications. In view of all these problems, it is felt that it will be better to use the available time and money on medium and small projects and on extensive water harvesting all over the country.

Other Controversial Issues

Apart from these specific issues, there is a difference of opinion regarding the overall approach to water management. As per the traditional concept of development, water is viewed as a resource but according to the environmental lobby, often it is overused under the name of development. In many areas, water tables are going down due to over-extraction of ground water. According to the principles adopted in the Rio Declaration on Environment and Development, nations need to ensure the equitable allocation of any resource within the present generation and between the present and future generations. Though in principle it is agreed, it is rather difficult to strike a nice balance between environmental perspectives and development.

Developmental Perspective

Though water is recognized as a free good and access to it as a human right, it tends to become a tradable commodity in scarcity situations. There is an apprehension that if this view (tradable good) gets strengthened in the context of privatization and globalization, forces of demand and supply will rule and the poor will be deprived of this basic necessity. Already, in many big cities, water is supplied by tankers on payment to better-off localities while the slums have to go without water for days. If the state does not take the appropriate measures to ensure adequate water supply to all, there is a fear that it will lead to corporatization of water and even MNCs will be attracted to capture the large and growing urban middle and upper middle class affluent market for water in India. India has to be vigilant and careful about the persistent advocacy of privatization of water services by the World Bank, Asian Development Bank, the Global Water Partnership and World Water Council (Iyer, 2004)

It is also debated whether bureaucratic control with all the rules and regulations can give the desired results for equitable distribution of water resources or it can be better achieved by treating water as a common property resource and by encouraging the community management of water like Pani Panchayat in Maharashtra. There is always a danger of powerful vested interests capturing the entire control, unless the state provides the safeguards the interests of the weaker sections.

It is interesting to see how far the existing water policy and the New Environment Policy have taken clear stand on these issues and provided necessary mechanism to ensure the effective implementation.

References

Abdul Kalam, A.P.J. (2005): "Integrated Water Mission" in *Science Information Notes*, July Special Issue on Water", New Delhi: Indian National Science Academy.

Ghosh, G.K. (2002): *Water of India: Quality and Quantity*. New Delhi: A.P.H. Publishing Corporation.

Government of India, Ministry of Environment and Forests (2004): "National Environment Policy – The Draft". New Delhi: Government of India.

International Institute for Population Sciences (IIPS) and ORC Macro (2000): *National Family Health Survey (NFHS-2): All India Report*, Mumbai, IIPS.

Iyer, Ramaswamy R. (2003): *Water: Perspectives, Issues, Concerns*, New Delhi: Sage Publications.

– (2005): "Towards Universal Guidelines on Water Use", *The Hindu*, Dec. 27.

Kumar, Rakesh, R.D. Singh and K.D. Sharma (2005): "Water Resources of India", in *Current Science* , 89 (5), September 10, pp.794-811.

National Commission for Integrated Water Resources Developmental Plan (1999): "Report of the Working Group".

Upadhyay, V. (2004): "National Environment Policy: A Critique of the Draft", *Economic and Political Weekly,* 39(39), September 25, p. 4306.

7 Towards the Need for Water Ethics for Sustainable Development

K. Radhakrishna Murty

The Concept of Sustainable Development

The notion of sustainable development is quite common place today in the parlance of discussion of developmental issues in the disciplines of population sciences, sociology and other social sciences. There is a consensus among the scientists that the central concern of development should be improvement in the state of human well-being of a society over time and that the natural environment, now – a scarce resource, has a determining role in such a state of affairs. The World Commission on Environment and Development defined sustainable development as a concept of inter-temporal resource use based on an ethical theory of inter-generational equity. To be more specific, the concept required the present generation to so use the natural, man-made, and human capital resources that they leave behind enough of these for the future generation to be able to reach at least the same level of social well-being as enjoyed by themselves.

As the world's population and per capita consumption grow, the human race is using resources and generating waste faster and faster. Thus, we are now beginning to see nature's limits. Scarcity and pollution of air and water, destruction of forests, diminishing grasslands and loss of fertile soil are becoming critical problems with serious consequences for health, quality of life, food production and even on the ability of the earth to support human life. The need to adapt human life simultaneously to the 'carrying capacity' of the earth's biological systems and to the limits of renewable energy sources will require a new social ethic. The essence of this new ethic is accommodation – the accommodation of human numbers and aspirations to the earth's resources and capacities. This new ethic must above all arrest the deterioration of man's relationship to nature. Otherwise, the current levels of consumption of resources may result in diminishing nature's future productivity not just exhausting current supply but also "stealing from our children". Therefore environmental protection and sound resource management are the essential parts of development planning. This principle is known as sustainable development that is, meeting peoples' current needs while preserving nature's productive capacity for the future (Brown, 1990).

Water as a Critical Factor for Sustainable Development

"Water is free and plentiful" is a myth. The reality is: "Water is often the most critical limiting factor to sustainable development". Water, "the liquid gold" is a sacred and precious resource for wealth and health of humanity. It is one of the life sustaining elements of nature and second only to 'air' for survival of mankind and livestock. It may appear strange that water is going to be scarcer and scarcer day-by-day. A worldwide water crisis is not unlikely in the near future if this resource is not properly conserved and utilized. The signals are already being received in many countries including India. The scarcity of water is mainly because of the mismatch between availability and demand, at times and in places both in terms of quality and quantity. The increase in population and consequent increase in water requirement for various uses (domestic, irrigation, hydro-power, industrial, etc.), the tapping of all easily available water resources, the delay in future project implementation due to increasing social and environmental awareness and increase inhuman developmental activities leading to contamination of available water resources are the prime causes for the scarcity of water. It is high time that this situation is realized and improved for the well-being of humanity.

At Global, Regional and Indian Perspective

It is said that if we look at the globe from outer space it looks like a water planet. However, while 70 per cent of the globe is covered with water, only three per cent of that is fresh water. Only one per cent of the land surface of the entire world is made up of fresh water ecosystems. And probably half the world's population live near those fresh water ecosystems (Hales, 1999). Take the case of Asia. With 60 per cent of the world population, it has only 36 per cent of global fresh water runoff and 80 per cent of that occurs in floods from May to October, exceeding man-made storage capacity and making it difficult to capture. Of the available fresh water in Asia, about 86 per cent use is for agriculture, 8 per cent for industry and 6 per cent for domestic purposes (Clarke, 1993). It is estimated that today at least half a billion people of the world live in regions with severe water shortages. By the year 2025, the number will explode fivefold to 2.8 billion people – about 35 per cent of the world's projected total of 8 billion people. Humans already use about 54 per cent of all accessible surface water runoff (usable and renewable fresh water). This is expected to increase to 70 per cent by the year 2025. Over the next two decades, population increase alone, not to mention growing demand per capita, is projected to push all the humanity into water scarcity (Hinrichsen et al., 1998).

Coming to India, of the total geographical area of about 328.8 million hectares (32,87,78,000) available in the country, about 33 per cent area is considered as drought affected area covering 13 major states of the country. The average annual surface flow in the river systems of India has been assessed to be 1953 km^3 (inclusive of 432 km^3 of groundwater), out of which the utilizable quantity will be about 1086 km^3 comprising 690 km^3 from surface sources and 396 km^3 from groundwater. The quantum of water available per capita in India is quite low as compared to many developed countries (17,500 cubic metre per year in USSR against 2,200 cubic metre per year in India). The utilizable surface water resources constitute to only 37 per cent of the annual flow in the river systems.

Thus, dwindling fresh water supplies per capita in the presence of rapid population growth are threatening the health and living standards of millions of people in a growing number of countries, as well as undermining agricultural productivity and industrial development. After all, water is the basis of life, and our stewardship of it will determine not only the quality of life but the staying power of human societies.

Efforts Made at Global Level to Solve Water Crisis

The first U.N. International Conference to draw attention to the coming water crisis at global level took place in the year 1977 at Mardal Plata, Argentina. Several others have followed including the Global Consultation on Safe Water and Sanitation for the 1990s held in New Delhi in 1990, and the International Conference on Water and the Environment, held in Dublin in 1992. The Dublin Water Principles, agreed to at the 1992 Conference, summarize the principles of sustainable water management:

Principle 1 Fresh water is finite and vulnerable resource, essential to sustain life, development and the environment.

Principle 2 Water development and management should be based on a participatory approach, involving users, planners, and policy makers at all levels.

Principle 3 Women play a central part in the provision, management, and safeguarding of water.

Principle 4 Water has an economic value in all its uses and should be recognized as an economic good.

More recently, in 1997, a comprehensive assessment of global freshwater resources based on a series of expert background analyses, was prepared for the fifth session of the U.N. Commission on Sustainable Development. As 1998 report of the Secretary-General states: "The assessment concludes that water shortages and pollution are causing widespread public health problems, limiting economic and agricultural development, and harming a wide range of ecosystems. Those problems may threaten global food supplies and lead to economic stagnation in many areas of the world. The result could be a series of local and regional water crises, with serious global implications".

Turning principles into practice will be difficult, however. Most countries need massive investments in sanitation and water supply infrastructure. In the developed world, for example, the United Kingdom must spend close to $ 60,000 million, building waste water treatment plants over the next decade in order to meet new European water quality standards. This amounts to about $ 1000 for every person in the country. Similarly, in developing countries, one of the most pressing problems is the overwhelming need to invest heavily in sanitation facilities and the provision of clean water. In Latin America alone, for instance, it is estimated that investments in water resources management and infrastructure will require $ 100,000 million over the course of the next two decades (Hinrichsen et al., 1998).

An important part of any international water management strategy is to help countries that share river basins, fashion workable policies to manage water resources more equitably. A water-short world is an inherently unstable world. Similarly, a water-short country is an inherently unstable country. Nearly 100 countries share just 13 major rivers and lakes. More than 200 river systems cross international borders. Naturally, conflicts can arise, especially where countries with rapidly growing populations and limited arable land collide over access to shared fresh water resources.

Towards the Need for Water Ethics

Then, why has so much of modern water management gone awry? And why is that even greater amounts of money and even more sophisticated engineering have not solved the world's water problems? The answer is: We are trying to meet insatiable demands by continuously expanding a supply that has limits, both ecological and economic. A host of nuts-and-bolts measures – from thrifty irrigation techniques and rain-water harvesting to water-saving plumbing fixtures and waste water recycling – can help us out of this predicament by reducing the amount of water required to grow food, produce material goods and meet household needs. But these measures will only spread rapidly if economic incentives and regulations are adopted that promote conservation and efficiency instead of waste and profligacy (Sandra Postel, 1996). We have been quick to assume rights to use water but slow to recognize obligations to preserve and protect it. We have lost a sense of respect for the wild river, for the complex workings of a wet land, for the intricate web of life that water supports. By and large, water has become strictly a resource to be dammed, diverted and drained for human consumption. As Thomas Fuller lamented: "We never know the worth of water till the well is dry". Grasping the connection between our own destiny and that of the water around us is integral to the challenge of meeting human needs while protecting the ecological functions that all life depends on.

For sure, better pricing and more open markets will assign water a higher value in its economic functions, and breed healthy competition that weeds out wasteful and unproductive uses. We also need a set of guidelines and responsibilities that stops us from chipping away at natural systems until nothing is left of their life sustaining functions, which the market place fails to value adequately. In short, we need a water ethic – a guide to rightly conduct in the face of complex decisions about natural systems we do not and cannot fully understand. The essence of such an ethic is to make the protection of water ecosystems a central goal in all that we do. The components of water ethic would be:

1. To avoid wastage and conspicuous consumption at all levels,
2. To impose better and rational pricing policy for water use in different sectors, and
3. To be aware of the fact that using water is not only our right but as a duty associated with preserving, rational using and sharing with others who are in need of it.

Living by such an ethic would mean using less, whenever we can and sharing what we have. It is about being good neighbours – as individuals, companies, communities, states, regions and nations. And it establishes norms of responsible behaviour against which to judge the actions of each global citizen. In practical terms, a water ethic is part of a sustainable development code that entails a wholly new approach to economic progress, one that harmonizes economic goals with ecological criteria. For individuals, a water ethic calls for an examination of lifestyles and consumption patterns with an eye toward reducing personal claims on the earth's finite water supply. Governments, the World Bank and other lending institutions that help set water priorities and directly fund many projects can help make this new ethic operational by placing ecological sustainability at the core of their investment policies and decisions. More often than not, this will favour smaller-scale projects: such as pumps and wells for ground water development, micro-catchments or small reservoirs to collect and store local runoff, over large dams and river diversions that tend toward greater disruption of natural systems. Ultimately, securing sufficient water for people while leaving enough for a healthy environment overall descends on rapid population growth. If current trends persist, per capita water supplies worldwide will drop by more than a third by 2025. Efforts to raise living standards for the additional 3 billion people will place enormous strains on natural systems. Moreover, human numbers continue to grow fastest in some of the most water-short regions.

Suggestions for Averting the Crisis

To avert such a crisis, at least in water-short countries, national governments need to give water resources management their highest priority. Crafting and implementing a national water strategy is essential to sustainable development. Such a strategy should include the guidelines such as:

1. Adopting a watershed or river basin management perspective, especially in water-short regions and initiating programmes on integrated development of river basins;
2. Instituting a workable water infrastructure so that national, regional and local water needs can be met within the context of a national water policy, and maintaining strict vigil in the implementation of various norms incorporated in such a policy;
3. Enacting and enforcing water legislation and regulations that conserve water and value the resource properly according to type of use;
4. Connecting water management to the needs of agriculture, industry, and municipalities, and meeting public health requirements for proper sanitation and disease prevention;
5. To create awareness and education among masses on water management and use;
6. To bring in the active role of non-governmental organizations in the field of water management, water engineering and water conservation; and
7. Water management and development should be based on a participatory approach involving users, planners and policy makers at all levels.

Watershed management refers to managing an entire land area served by all the rivers and aquifers that drain into a particular body of water. Deforestation is, of course, causing a lot of damage to such watershed management programmes wherever planned. Also rainwater runs off so fast that little can be collected for use or savings. Tampering with watersheds has proved ruinous for many developing countries, where hillsides denuded of vegetation empty tonnes of soil into water courses every year, causing floods, during the wet seasons and suffocating aquatic life during the dry seasons. In our country, as a result of the 1987 National Water Policy, the states of Rajasthan and Gujarat have set up a committee to regulate and control water use in the Sabarmati river basin which encompasses parts of both states. Basing on the performance of such a system, it may be extended to other water-short areas of the country with high population densities.

Managing watersheds and river basin's sustainability also means building institutional capacity, including the creation of cross-sectional data collection and monitoring systems. Capacity-building should be a key theme of international and national organizations promoting water management schemes. Added to this, fresh water must be valued to reflect its status as a scarce resource, instead of being treated as a free or nearly free resource. As the Organization for Economic Cooperation and Development (OECD) points out, proper pricing policies can encourage environmentally responsible water-use behaviour, as well as help to assure an adequate supply of water. To accomplish this, water should be valued appropriately in each of its various uses. The introduction of water markets and pricing mechanisms can have immediate and lasting impacts on water use. A World Bank study of the water market system in Chile concluded that it contributed greatly to better management and fairer pricing.

Further, a workable water management system requires the institutional capacity to balance sectoral needs for the good of society as a whole and also to consider ecosystem needs. Water allocation, rather than absolute scarcity of water, often lies at the heart of national water problems. In developing countries like ours, meeting sectoral demands is challenging because most lack efficient water management systems and equitable pricing policies that are based on how water resources are used. Without policies that link the supply of freshwater to competing sectoral uses, local and regional water shortages often result, and competition becomes increasingly bitter.

The world needs sustainable water management, but we are not headed in the right direction fast enough. Without moving in a new direction, many more areas will face water shortages, many more people will suffer, more conflicts over water will occur, and more precious wetland ecosystems will be destroyed. While a freshwater crisis appears inevitable in many water-short regions, in others the problem could be managed if appropriate policies and strategies were formulated, agreed to, and acted on soon. A vital part of a long-term solution is world-wide recognition of the links between rapidly growing populations and shrinking freshwater supplies. Recognition, knowledge, and concern can help build the political will to avert a crisis and develop the commitment needed to assure that humanity's apparently unquenchable thirst for fresh water does not exhaust the world's finite water supply.

"India too faces a turbulent water future. Unless dramatic changes are made and made soon, the way in which the government manages water, India will have neither the cash to maintain and build a new infrastructure nor the water required for the economy and the people", according to a World Bank report (Gargi Parsai, *The Hindu*, October 2, 2005). The country's development of water infrastructure is not accompanied by an improvement in governance of water resources and water services. "The Indian state water apparatus still shows little interest in the key issues of the management stage: participation, incentives, water entitlements, transparency, entry of the private sector, competition, accountability, financing and environmental quality", says the World Bank report prepared by John Briscoe. Making a strong case for the construction of mega-dams, the report says "that India can store only about 30 days of rainfall compared to 900 days in major river basins in arid areas of developed countries. The decline in the quality of public irrigation and water supply services will normally be expected to produce social unrest and political pressure. The economic ideas of the 1991 economic reforms must be drilled down from the regulatory and financial sectors into the real sectors including water sector if India is to have sustainable economic growth". Evidence abounds of the inability of the state water machinery to address even problems of the provision of public irrigation and water supply services. User charges are negligible resulting in lack of accountability and insufficient generation of revenue even for operations and maintenance. The gap between tariff and value of irrigation and water supply services has fuelled endemic corruption. Therefore, the role of the state must change from that of builder and controller to creator of an enabling environment of the actions of large and small water users. It is said "one can fool some people all the time and all the people for some time but one cannot fool all the people all the time". The link proposals and the renewed activity now undertaken by National Water Development Agency (NWDA) to connect all the major rivers in India under the Ministry of Water Resources appear to be trying to fool all the people all the time and succeeding too. These link proposals which started very seriously with all sincerity, lost direction mid way through and became a farce. Now, they appear to be heading towards a tragedy, a Greek tragedy at that (Reddy, 2003).

Conclusion

Falling freely from the sky, water has deluded us into believing it is abundant, inexhaustible and immune to harm. The challenge now is to put as much human ingenuity into learning to live in balance with water as we have put into controlling and manipulating it. As Sandra

Postel (1996) rightly pointed: "The last oasis of conservation, efficiency, recycling, and reuse is large enough to get us through many of the shortages on the horizon, buying us time to develop a new relationship with water systems and to bring consumption and population growth down to sustainable levels". An American-Indian tribal saying sums up thus: "Only when the last tree has been felled, only when the last river has been poisoned, only when the last fish has been caught, only when you will learn that money cannot be eaten".

Sustainable development is all about restraining the foolish man from committing these foolish acts. Yet, the pace of this transition needs to quicken if we are to avert severe ecological change, economic setbacks, food shortages and regional, national and international conflicts. In the end, the time available to adjust may prove as precious as water itself. The world needs another blue revolution in water management, just as we need another green revolution in agriculture. Achieving such a blue revolution will require coordinated policies and responses to problems at international, national and local levels.

References

Brown, J.W. (1990): "U.S. Policy in the Crucial Decade Ahead", in J.W. Brown (ed.), *In the U.S. Interest*, West View Press, Boulder, Colorado.

Clarke, R. (1993): *Water: The International Crisis*, M.I.T. Press, Boston, Massachusetts.

Gargi Parsai (2005): "India Faces a Turbulent Water Future: World Bank", *The Hindu,* 2 Oct., 2005.

Hales, D.F. (1999): "Fresh Water: Will the World's Future Needs be Met", in *Troubled Waters: Managing our Vital Resources*, Global Issues, USIS August Issue.

Hinrichsen, D., et al., (1998): "Toward A Blue Revolution", *Population Reports of the Johns Hopkins University School of Public Health*, Baltimore, MD, September Issue.

Reddy, M.S. (2003): "Linking of Rivers in India: Retrospect and Prospect", in *Journal of Applied Hydrology,* Vol. XVI, No. 4A, October 2003.

Sandra Postel (1996): "Water Ethics", Rashmi Mayur (ed.) in *Earth, Man and Future,* International Institute for Sustainable Future, Mumbai.

8 Water Consumption Patterns in Domestic Households in Major Indian Cities

Abdul Shaban and R.N. Sharma

Predicting future is a hazardous business but one thing is certain that the world, at least in some respects, in future is going to be different from that of today (Figureres et al., 2005). Among others, the demographic transition, geographical shift (urban to rural, and under-developed to developed areas) of population, technological advancement, increased extent and impact of globalization, degradation of environment and emergence of water scarcity due to increased demand, can be predicted with more certainty that the societies are going to confront in future. Water, the need of life, is possibly to pose greatest challenge on account of its increased demand with population rise and economic development, and shrinking supplies due to over exploitation and pollution. Although, water is a renewable natural resource abundant on globe, as about two-thirds of the planet is covered with this, the fresh water is limited, as only 2.7 per cent of the total water on globe is the fresh water. In fact, a very small proportion of this 2.7 per cent is effectively available for human use. The availability of fresh water varies greatly with space and time, and increased scarcity has made it an economic good and therefore access to it also varies with socio-economic status of population. The growth process and expansion of economic activities inevitably lead to increasing demand for water, and in a country like India as the development trickles down to rural areas, the demand for water will increase in rural areas as well. The sharing and command on water possibly may create increased tension and disputes between rural and urban areas. The emerging scarcity of water has raised a host of issues related to sustainability of present kind of economic development, sustanious water supply, equity and social justice, water financing, pricing, governance and management. The present study mainly focuses on domestic use of water in seven major Indian cities, namely, Delhi, Kanpur, Kolkata, Ahmedabad, Mumbai, Hyderabad and Madurai. Besides focusing on activity-wise and socio-economic group-wise consumption of water, the paper also attempts to find out sources and perception of households about safety and wastage of water, duration of municipal water supply, and various measures adopted by the households for conservation of water. The organization of the rest of the study is as follows: Section II mainly deals with methods used for data collection and analysis, while section III discusses the recommendations of various agencies about the per capita need of water in urban areas. Exploration and analysis of socio-economic groups, area and city-wise domestic consumption of water are presented in section IV. Section V extends the analysis presented in section IV, by analysing activity-wise consumptions. Sources of water and perceptions of households about the safety of water are analysed in section VI, while section VII deals with duration and frequency of supply of municipal water and distance of sources of water. The shortage and unreliability of water supply in urban India has led to storage of water by households to meet their needs. Section VIII examines the extent and magnitude of the water storing practices adopted by urban households. Household perception about the wastage of water, and the methods adopted by households for conservation of water are explored in sections IX and X, respectively. The last section presents summary and conclusions.

Data and Methodology

The household survey, to find out the consumption, availability, access, storage, and methods adopted for conservation of water at domestic households in seven major Indian cities, Delhi, Greater Mumbai, Kolkata, Hyderabad, Kanpur, Ahmedabad and Madurai, was conducted in March 2005. As per the Census of India 2001, these cities, except Madurai (municipal corporation) are among 27 municipal corporations with million-plus population, and their respective populations are 9.8, 11.9, 4.6, 3.5, 2.5, 3.5 and 0.92 million, respectively. While selecting the cities for the study, adequate regional presentation was kept in mind, and therefore two cities each from northern (Delhi, Kanpur), western (Mumbai, Ahmedabad), southern (Ahmedabad, Madurai) regions, and one from the eastern region (Kolkata) were selected. The number of sample households selected from each of these cities were largely based on the size of their populations. Total 2,734 households were covered by the survey in these seven cities, and the city-wise distribution of households was: 507 in Delhi, 500 in Mumbai, 400 in Kolkata, 398 in Hyderabad, 303 in Kanpur, 361 in Ahmedabad, and 265 in Madurai. Each of these cities were categorized into five different areas: (i) high income group (HIG) areas with well planned buildings, (ii) middle income group (MIG) areas with well planned buildings, (iii) low income group (LIG) areas with well planned buildings, (iv) slum areas, and (v) the mixed areas. Within these identified clusters, random sampling of households from electoral role was done. In each of these cities, several clusters of the same kinds of areas can be found, and therefore, from each cluster at least 8 interviews were conducted. The data was collected through a structured schedule and target respondents were the housewives. The volume of vessels in which households used to store water was measured and the number of vessels of water used in different activities was ascertained. Where running tap or piped water was used in some activities, the duration for which the tap remained open was arrived at and the quantity of water per minute coming out from the tap was measured. By multiplying the time with the quantity of water per minute, the amount of water used through running taps was arrived at. The amount of water used in toilet was assessed by volume of bucket used, and flush tank capacity. The assessment of quantity of water used in various activities was a difficult task, but all the efforts were put in, to assure the accuracy of data.

Along with the consumption of water, we also collected data on assets possessed by households. As the income data is generally considered unreliable, we have used asset scores to classify households in various economic classes, such as the 'very poor', 'poor', 'lower', 'middle' and 'upper' classes. We assigned certain weights to the assets possessed by households and by summing up the scores for each asset, 'asset score' for a household was arrived at. The assets for which data were collected and weights assigned to them, are as follows: four-wheeler/car (weight 5.0), refrigerator (2.0), washing machine (2.0), microwave oven (2.0) three/two-wheeler (1.5), coloured television (1.5), computer/laptop (1.0), musical devices like cd/dvd/vcd/MP3 player (0.5), internet connection (0.5), black and white television (0.5), mobile phone (0.5), cable for television (0.25), and radio/tape recorder (0.25). The following ranges of asset scores were used to classify the households in the above mentioned economic/asset classes:

1. Very poor class: asset score 0.00 (has no assets on which data is collected).
2. Poor class: asset score between 0.01 to 0.99 (at least has radio/tape recorder or black and white/colour television).
3. Lower class: asset score between 1.00 to 4.99 (black and white television, radio/tape recorder, cable, mobile phone, or some of them with other assets).
4. Middle class: asset score between 5.00 to 9.99 (at least has coloured television, refrigerator, washing machine or some of them with other assets).
5. Upper class: asset score 10.00 and above (at least has four-wheeler, refrigerator, washing machine, and colour television, or some of them with other assets).

Annual income data of households was also collected. The Pearson Correlation Coefficient between the household asset score and the income was found to be 0.722 (p-value 0.0001). However, we have not used annual household income data, but the assets score, for classifying households in various economic categories, as there is a greater possibility of under or over reporting of the income. Besides categorizing households on the basis of asset scores, we have also classified the households in five socio-economic classes (SEC) based on education and occupation of the main wage earner of the household. These five classes are SEC-A, SEC-B, SEC-C, SEC-D and SEC-E.

Quantity of Water Required for Different Activities

A number of factors like climate, culture, food habits, work and working conditions, level and type of development, and physiology determine the requirement of water. As per the Bureau of Indian Standard, IS:1172-1993, minimum water supply of 200 litres per capita per day (lpcd) should be provided for domestic consumption. IS:1172-1993 also mentions that the amount of water supply may be reduced to 135 lpcd for the LIG and the economically weaker sections (EWS) of the society. As such, for small towns, a minimum water supply of 135 lpcd should be provided for domestic consumption (Modi, 1998). The break-up of domestic water requirements for cities with full flushing system and small towns, given by IS:1172-1993, is reproduced in Table 1.

Table 1

Domestic Water Requirement in Indian Towns and Cities

Use of Water	*Cities/Towns with Full Flushing System (lpcd)*	*Small Towns (lpcd)*
Drinking	5	5
Cooking	5	5
Bathing	75	55
Washing clothes	25	20
Washing utensils	15	10
Washing and cleaning of house and residence	15	10
Lawn watering and gardening	15	-
Flushing of water closets, etc.	45	30
Total	200	135

lpcd: litres per capita per day.

Source: Modi (1998).

Besides domestic requirement, water is also in demand for commercial, industrial, and civic or public use. The IS:1172-1993 gives the requirement for these three sectors as 20 lpcd, 50 lpcd, and 10 lpcd, respectively. Thus, the total requirement of water in industrial and commercial towns, with full-flushing system, becomes 280 lpcd. The concerned documents of Ninth Plan (1997-2002) have advocated the requirements of water in urban areas as 125 lpcd in cities, with the planned sewerage systems; 70 lpcd in cities without planned sewerage system; and 40 lpcd for those collecting water from public stand posts. However, under the Tenth Plan (2002-07), the cities with planned sewerage system were classified into two groups based on population, i.e., metropolitan or megacities and non-metropolitan cities. In the former, the recommended minimum water supply levels is 150 lpcd and in the latter 135 lpcd. (Government of India, 1997, 2002). The National Commission on Urbanization (1988) recommended a per capita supply of water of 90-100 litres per day, needed to lead a hygienic existence, and this level of supply must be ensured to all citizens (Ramachandraiah, 2001).

Notwithstanding the IS:1172-1993 and the Five-Year Plan recommendations, we find that almost every municipal corporation/ municipality has defined the requirement of water per capita per day in its own way. One agrees that industrial and commercial development of towns and cities may differ and so the required amount of water, but requirement for domestic use need not vary so much. The Municipal Corporation of Greater Mumbai (MCGM) advocates 135 lpcd as the domestic requirement of water, but Delhi Development Authority (DDA) considers 225 lpcd per day as the needed water for domestic use. The DDA assess water requirement of another 75 lpcd for industrial, commercial, and civic or public use, thus putting the total requirement of water in Delhi as 300 lpcd. This wide variation in recommendations/prescriptions for domestic use of water seems inexplicable, particularly when both the megacities have well-developed sewerage/flushing systems. Perhaps, the supply constraint in Mumbai has led the Municipal Corporation to assess the lower requirement of water for the domestic use.

The World Health Organization (WHO) classifies the supply and access to water in four service categories. These categories are: (1) no access (water available below 5 lpcd), (2) basic access (average, approximately 20 lpcd), (3) intermediate access (average, approximately 50 lpcd), and (4) optimal access (average of 100-200 lpcd) (WHO, 2003; see also Bartram, 2003). Considering the fact that various agencies recommend various quantities of requirement of water for domestic use, we have taken 100 lpcd consumption (a sign of availability, as consumption is determined by the availability) of water as benchmark for finding out water deficient households. It must be noted here that there is no strong basis for this 100 litres benchmark but it is a some kind of average requirement one must get, in order to live with minimum health and hygiene.

Domestic Consumption of Water

It is important to note here that the quantity of water consumed in most of the Indian cities is not determined by the demand but the supply. People attempt to adjust to the quantity (as well as quality) of water supplied. The 54th Round data of National Sample Survey Organization (NSSO) reveals that 80 per cent of the households in urban India, across different segments, consider that they have sufficient water supply (Bajpai and Bhandari, 2001). In fact, this shows nothing but adjustment of people to the supply so much that they do not feel that more water is required. This in turn creates hygiene and sanitation problems resulting in several health consequences.

Table 2

Domestic Water Consumption per Household and Per Capita Per Day

(in litres)

Cities	*Per Household*		*Per Capita*	
	Mean	*Std. Deviation*	*Mean*	*Std. Deviation*
Delhi	377.7	256.8	78.0	49.9
Mumbai	406.8	158.6	90.4	32.6
Kolkata	443.2	233.6	115.6	64.9
Hyderabad	391.8	172.0	96.2	43.8
Kanpur	383.7	286.2	77.1	58.2
Ahmedabad	410.9	224.1	95.0	54.6
Madurai	363.1	182.1	88.2	44.4
Total	398.3	220.20	91.56	51.51

Table 2 shows per household as well as per capita consumption of water in seven major cities in the country. It is very obvious from the Table that in all the cities, the consumption (indication of availability) of water per capita is much lower than what is recommended by the Bureau of Indian Standard, IS:1722-1993, and the Tenth Plan document. Moreover, it is even lower than the recommended level for LIG colonies and weaker section households. The data is also an indication for the lower public hygiene and sanitation conditions in Indian cities. The average per capita water consumption in domestic households for all the seven cities is about 92 lpcd. The highest consumption is in Kolkata (116 lpcd), followed by Hyderabad (96 lpcd), Ahmedabad (95 lpcd), Mumbai (90 lpcd), Madurai (88 lpcd), Delhi (78 lpcd), and Kanpur (77 lpcd). It is surprising to find that in Delhi, water consumption is so low when Delhi Jal Board claims supplying, on an average, 211 lpcd per household. Similarly, the Brihan Mumbai Municipal Corporation (BMC) claims of supplying 198 lpcd (total water supply to the city, 2950 million litres per day – less 20% lossess), but the consumption in Mumbai, as mentioned above, is only about 92 lpcd. In terms of international comparison, the Indian cities have far less consumption of water. For example,

Figure 1
Distributional Characteristics of Per Capita Water Consumption (in litres) in Major Indian Cities

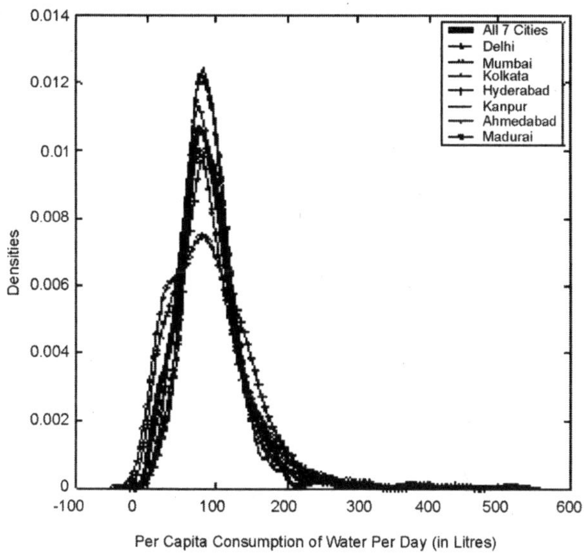

domestic water consumption in Munich is 130 lpcd, and in Amsterdam it is 156 lpcd. In Singapore, Hong Kong, Sydney and Tokyo, the consumptions are 162 lpcd, 203 lpcd, 254 lpcd, and 268 lpcd, respectively (*Down to Earth*, 2005).

Overall, in terms of per capita consumption of water, the condition in two North Indian cities, Delhi and Kanpur, seems to be worse. The dispersion statistics (standard deviation) shows that wide variations in per capita consumption of water also exist in North Indian cities. Besides, having the highest level of water consumption, Kolkata also has wide variation in consumption of water per capita. In terms of the variation, Kanpur, Ahmedabad and Delhi follow Kolkata. The distributional characteristic of per capita water consumption is further illustrated with the help of normal 'kernel density plot' (Figure 1). In case of Delhi, a small ridge is visible, besides major nodal peak, at around 30-50 litre per capita water consumption. This shows that a sizeable chunk of population in Delhi consumes water only about 30-40 lpcd. Besides Delhi, Kolkata also has small ridges at around 175 lpcd. The distribution in rest of the cities is smooth. However, a long

Figure 2

Distributional Characteristic of Household Asset-Score in Indian Cities

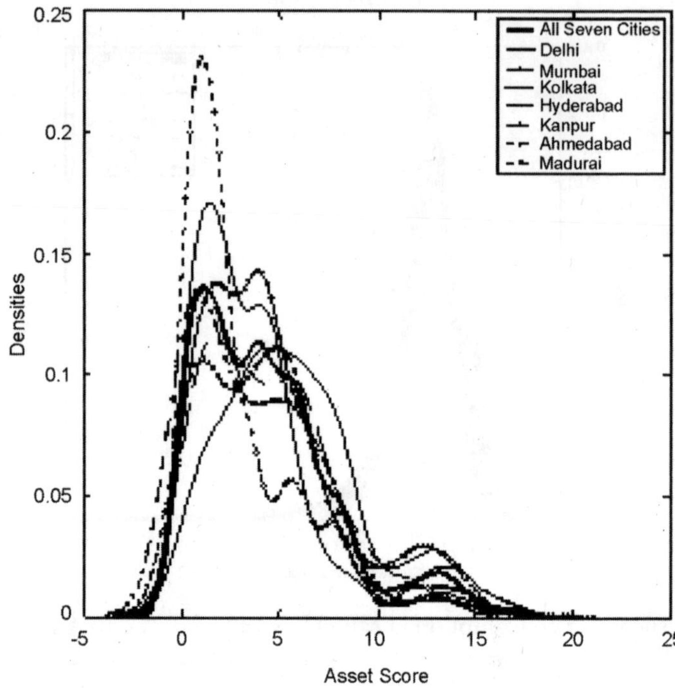

positive skewness of per capita water consumption is visible in almost all the cities.

It is not that the consumption is lower only in LIG areas and slums in Indian cities, but also in HIG and MIG areas the consumption is constrained, and thus below expectation. Table 3 shows that, whereas slum areas in the seven selected cities consume, on an average, about 82 lpcd, it is only about 100 lpcd and 94 lpcd in the case of HIG and MIG areas, respectively. LIG areas consume only about 90 lpcd. Figure 2 shows distributional characteristics of 'household asset score' with the help of 'normal kernal density'. The figure demonstrates that there are wide economic differences among households in Indian cities. However, in terms of consumption of water, the differences are not so extreme. This further indicates that the consumption of water in Indian cities is related to supply, rather than demand constraint.

Table 3

Area-wise Consumption of Water Per Household and Per Capita Per Day

(in litres)

Area	*Per Household*		*Per Capita*		*N*
	Mean	*Std. Deviation*	*Mean*	*Std. Deviation*	
High income group areas with well planned building	402.5	230.3	99.9	59.8	551.0
Middle income group areas with well planned building	396.4	248.6	94.2	57.6	571.0
Low income group areas with well planned building	393.5	176.4	90.2	40.6	552.0
Slum areas	398.7	216.8	81.9	41.1	530.0
Others (mixed areas)	400.5	222.0	91.3	53.1	530.0
Total	398.3	220.2	91.6	51.5	2734.0

Further, the analysis reveals that although there are socio-economic class-wise variations in consumption of water, the variations are not very significant (Table 4). Thus, the supply side constraints have been able to maintain a high equality in consumption of water in Indian cities. For example, whereas the SEC-E consumes 78.9 lpcd, it is 102.1 and 95.2 lpcds for the SEC-A, and SEC-B, respectively.

Table 4

Socio-economic Class-wise Consumption of Water Per Household and Per Capita Per Day

(in litres)

Socio-economic class	*Per Household*		*Per Capita*		*N*
	Mean	*Std. Deviation*	*Mean*	*Std. Deviation*	
SEC–A	407.1	233.3	102.1	62.8	639.0
SEC–B	399.2	239.0	95.2	56.7	684.0
SEC–C	399.9	213.9	88.4	42.0	619.0
SEC–D	390.0	198.7	84.9	41.1	461.0
SEC–E	387.9	192.5	78.9	39.3	331.0
Total	398.3	220.2	91.6	51.5	2734.0

Table 5

Water Consumption Category-wise Distribution of Households

(%)

Litres per capita per day	*All 7 Cities*	*Cities*						
		Delhi	*Mumbai*	*Kolkata*	*Hydera-bad*	*Kanpur*	*Ahmeda-bad*	*Madurai*
Below 50	17.5	29.8	5.4	11.8	9.0	33.7	19.4	17.0
50 to 75	22.6	22.1	29.4	13.5	21.6	24.4	20.8	26.4
75 to 100	25.3	20.7	34.2	22.0	32.4	17.2	21.9	26.0
100 to 135	20.2	17.2	23.6	24.3	23.4	14.2	18.8	17.4
135 to 175	8.7	5.3	5.0	15.0	8.3	6.9	12.7	10.2
175 to 200	2.2	2.2	1.4	4.8	2.3	1.3	2.2	.8
Above 200	3.4	2.8	1.0	8.8	3.0	2.3	4.2	2.3
Total	100.0	100.0	100.0	100.0	100.0	100.0	100.0	100.0

In these seven cities, a very small proportion of population consumes water above 100 litres per capita per day. As shown in Table 5, only about 35 per cent of the total population in the cities consumes water above 100 lpcd. There also exist wide variations in proportion of population in different cities consuming water above 100 lpcd. For example, in Kolkata 52.9 per cent of the households consume water above 100 lpcd, but in Kanpur only 24.7 per cent consume water above 100 lpcd. Likewise, Ahmedabad (37.9% of the households), Hyderabad (37%), Mumbai (31%), and Madurai (30.7%) follow Kolkata in terms of the consumption of water above 100 lpcd.

Kanpur has a dubious distinction among the cities under focus. In this city more than one-third of population uses water below 50 lpcd. In this respect, Kanpur is followed by Ahmedabad (19.4% of population consuming water below 50 lpcd), Madurai (17.0%), Kolkata (11.8%), Hyderabad (9.0%), and Mumbai (5.4%).

As has earlier been stated, not only in LIG and slum areas of the Indian cities but also in HIG and MIG areas, the consumption of water is low. The Table 6 shows interesting results that the water consumption among different areas in a city is not very different from each other, except that the model class of consumption for HIG, MIG and LIG areas is formed by the consumption class of 75-100 lpcd, and in the case of slums, and mixed areas, it is 50-75 lpcd. However,

Table 6

Area and Consumption Category-wise Distribution of Household (%)

Litres per day per capita	*Area*				
	High income group area with well planned building	*Middle income group area with well planned building*	*Low income group area with well planned building*	*Slum area*	*Others (a mixed area)*
Below 50	19.8	17.7	13.2	21.3	15.5
50 to 75	16.3	20.5	23.4	27.0	26.2
75 to 100	21.2	26.4	31.0	24.2	23.8
100 to 135	19.2	20.0	21.4	17.9	22.5
135 to 175	12.9	8.4	7.6	6.2	8.5
175 to 200	3.8	2.1	1.4	2.3	1.3
Above 200	6.7	4.9	2.0	1.1	2.3
Total	100.0	100.0	100.0	100.0	100.0

whereas 23.4 per cent of the population in HIG areas uses water above 135 lpcd, in slum areas only 9.6 per cent of population uses water above 135 lpcd. The consumption of water by various socio-economic classes shows that although a sizeable proportion of households in all the classes consume water below 50 lpcd, it is in the case of SEC-E where the consumption is very low. More than one-quarter of the population in this class consumes water below 50 lpcd (Table 7). Also, as compared to 23.3 per cent of the SEC-A consuming above 135 lpcd, it is only 8.1 per cent of the population in SEC-E, which consumes water above this limit. The consumptions by socio-economic classes do largely correspond with the consumptions based on the asset classes. Whereas 25.7 per cent of the population in SEC-E has the consumption below 50 litres per capita per day, 28.1 per cent of the population in 'very poor' category consumes water below this limit (Table 8). Similar to HIG, MIG and the LIG areas, the model class of consumption for 'upper', 'middle' and 'lower' categories of households, based on the asset score, is also 75-100 litres per capita per day (Table 8).

Table 7

Socio-economic and Consumption Category-wise Distribution of Household

(%)

Litres per capita per day	*Socio-economic Category*				
	SEC-A	*SEC-B*	*SEC-C*	*SEC-D*	*SEC-E*
Below 50	18.0	17.7	14.1	15.2	25.7
50 to 75	17.7	19.3	25.5	27.5	26.6
75 to 100	22.2	24.9	27.8	29.1	22.7
100 to 135	18.8	21.9	21.6	20.0	16.9
135 to 175	13.1	9.6	7.6	5.4	5.1
175 to 200	2.8	3.4	1.1	1.1	2.1
Above 200	7.4	3.2	2.3	1.7	.9
Total	100.0	100.0	100.0	100.0	100.0

Table 8

Asset and Consumption Category-wise Distribution of Household

(%)

Litres per capita per day	*Asset Class*				
	Very Poor	*Poor*	*Lower*	*Middle*	*Upper*
Below 50	28.1	16.7	13.1	21.5	21.7
50 to 75	23.1	25.9	25.3	18.8	15.8
75 to 100	24.0	28.4	25.9	24.1	22.2
100 to 135	15.7	19.5	22.6	18.8	16.3
135 to 175	5.0	6.2	8.6	9.5	13.1
175 to 200	1.7	1.7	1.9	2.6	3.6
Above 200	2.5	1.5	2.6	4.8	7.2
Total	100.0	100.0	100.0	100.0	100.0

Inadequate water supply in Indian cities seems to be a rule rather than an exception. Even if we take 100 litres per capita per day as the criterion for defining water deficient and sufficient households, 65 per cent of the sample households remain water deficient. The proportion of deficient households is the highest in Kanpur (75.2%), followed by Delhi (72.6%). In fact, except Kolkata, in all the other cities, the water deficient households are above 60 per cent. Area-wise classification of water deficient households (Table 9) shows that, as expected, these are slum areas which have maximum percentage of water deficient households in all the cities. In Ahmedabad and Kanpur cities, the percentages of water deficient slum households are as high as 86.1 per cent, and 82.1 per cent, respectively, while these range between 70 per cent to 75 per cent in the case of Delhi, Mumbai, and Hyderabad. Among the cities, the least percentage of water deficient households are found in Kolkata (47.3%). Areas, asset-classes and SEC classes of households are highly correlated with each other, as very similar percentages of water deficient households, based on each of the above criterion (see Tables 6, 7, and 8).

Table 9

Percentage of Water Deficient Households in Different Areas (consuming below 100 lpcd)

Areas	*All 7 Cities*	*Delhi*	*Mumbai*	*Kolkata*	*Hyderabad*	*Kanpur*	*Ahmedabad*	*Madurai*
HIG	57.3	68.4	48.8	39.8	64.6	75.0	42.5	66.7
MIG	64.6	78.1	69.2	42.1	61.7	66.1	56.3	76.8
LIG	67.4	71.7	76.2	55.7	61.7	70.2	69.2	66.7
Slum	72.5	72.4	74.1	58.1	65.4	82.1	86.1	70.0
Mixed	65.5	72.0	73.1	40.5	62.0	82.8	57.8	67.3
Total	65.4	72.6	69.0	47.3	63.1	75.2	62.0	69.4

Activity-wise Consumption of Water

At household level, bathing consumes highest amount of water. Together, in all the seven cities, it consumes about 28 per cent of the total water at household level (Table 10). Consumption in toilets (20%), washing clothes (18.6%) and washing utensils (16.3%) follow the consumption in bathing activity. On an average, less than 10 per cent of the total water in a household is used for drinking and cooking.

Among the cities, households (37.1%) in Kolkata consume highest proportion of total water in bathing. Delhi (31.7%) Kanpur (29.1%), Madurai (26.6%) and Hyderabad (25.6%) follow Kolkata in this regard. Toilet consumes 15-26 per cent of the water at household level in these cities, the highest being in Madurai (25.7%) and least in Kolkata (15.9%). Washing clothes also accounts for about 14-22 per cent of the total water consumption at household level. It is the highest in Mumbai

(24.3%), followed by Hyderabad (20.9%), and Ahmedabad (21.4%). Interestingly, North Indian cities Delhi, Kanpur and even Kolkata use less proportion of water in washing clothes than other cities. In these North Indian cities, the consumption at household level in washing clothes varies between about 14-16 per cent of the total water consumption in the respective cities.

Table 10

Activity-wise Distribution of Water Consumption in Cities (% of total consumption by households/day)

Activity	*All 7 cities*	*Delhi*	*Mumbai*	*Kolkata*	*Hydera-bad*	*Kanpur*	*Ahmeda-bad*	*Madurai*
Bathing	28.2	31.7	23.7	37.1	25.6	29.1	22.8	26.6
Washing clothes	18.6	14.2	24.3	14.0	20.9	16.3	21.4	18.9
Drinking	4.2	5.0	4.2	2.6	4.3	3.8	4.9	4.9
Cooking	3.0	3.7	1.7	2.3	3.1	3.2	3.3	4.2
Toilets	20.0	16.5	21.6	15.9	24.1	20.1	19.1	25.7
Cleaning house	7.3	7.0	6.6	11.7	3.5	5.7	12.4	1.9
Washing utensils	16.3	16.5	17.4	16.1	16.5	15.4	15.2	16.1
Others	2.4	5.6	0.5	0.3	2.0	6.3	0.9	1.7
Total	100.0	100.0	100.0	100.0	100.0	100.0	100.0	100.0

Table 11

Activity-wise Consumption of Water at Household Level

(%)

Activity	*SEC-A*	*SEC-B*	*SEC-C*	*SEC-D*	*SEC-E*
Bathing	29.0	26.4	28.2	28.9	29.9
Washing clothes	19.6	19.3	18.3	17.4	17.3
Drinking	3.9	4.2	4.3	4.4	4.5
Cooking	2.9	3.1	3.0	3.0	2.9
Toilets	17.0	19.0	21.6	22.4	21.9
Cleaning house	8.0	8.1	6.7	6.7	6.0
Washing utensils	16.4	16.4	16.2	16.0	16.5
Others	3.2	3.6	1.7	1.2	1.0
Total	100.0	100.0	100.0	100.0	100.0

Table 11 shows the activity-wise (%) share of water consumption for various SECs in the seven selected cities. In all the SECs, the consumption for bathing is very prominent. Of the total water consumed at household level, about 26-30 per cent in all the SECs is being consumed in bathing, followed by use in toilet (17-22%) and washing cloths (17-20%). Washing utensils accounts for about 16 per cent of the water consumed in households in almost all the SECs. Almost, similar results are observed when consumption pattern in various activities at household level is categorized based on the asset-score (Table 12).

Table 12

Asset Category and Activity-wise Daily Consumption of Water at Household Level

(%)

Activity	*Very Poor*	*Poor*	*Weaker*	*Middle*	*Upper*
Bathing	28.9	28.1	28.4	28.1	28.6
Washing clothes	17.8	18.4	17.9	19.5	19.7
Drinking	4.8	4.5	4.1	4.2	4.2
Cooking	3.7	3.2	2.8	3.0	3.0
Toilets	19.4	21.8	21.4	19.2	13.2
Cleaning house	8.0	6.3	7.4	7.0	9.2
Washing utensils	16.7	16.9	16.5	15.4	17.0
Others	0.6	0.9	1.5	3.7	5.2
Total	100.0	100.0	100.0	100.0	100.0

Sources of, and Perception about Safety of Water

A majority of households in urban India depend upon the municipal water supply for their daily needs. The 54th Round data of NSSO shows that 70.1 per cent of the households in urban India were dependent on the tap water (municipal supply), 21.4 per cent on tubewells, 6.7 per cent on wells, and the rest on other sources (Bajpai and Bhandari, 2001). However, our survey shows that as high as 92 per cent of the households in the seven major Indian cities under present focus are dependent on municipal water supply (tap water). Of this 92 per cent of the population, 9.5 per cent are dependent on community taps and the rest (90.5%) on their own private taps provided by their municipalities (Table 13). The proportion of households using

Table 13

Dependence of Household on Various Sources of Water (multi-sources possible)

(%)

	Main Source	*Sub-Source (% Contribution to Main Source)*	*All 7 cities*	*Delhi*	*Mumbai*	*Kolkata*	*Hydera-bad*	*Kanpur*	*Ahme-dabd*	*Mad-urai*
1	Tap water		92.0	91.9	98.6	98.8	99.7	54.1	94.2	98.5
1.1	(of which)	Private Tap	90.5	91.9	98.5	98.1	99.7	54.1	80.0	98.5
1.2		Community Tap	9.5	8.1	1.5	1.9	0.3	45.9	20.0	1.5
2	Tube well		38.1	24.5	3.6	27.8	48.5	79.2	38.5	81.1
2.1	(of which)	Private Tube well	24.0	6.3	3.6	1.0	41.8	36.4	37.2	76.7
2.2		Community Tube well	76.0	93.7	96.4	99.0	58.2	63.6	62.8	23.3
3	Tankers		2.1	2.0		0.5	10.6			1.1
3.1	(of which)	Municipal Tanker	59.3	60.0		100.0	54.5			100.0
3.2		Private Tankers	40.7	40.0			45.5			
4	Well/open well		1.0	0.2	0.6	2.8	3.3			
5	River/Canal/Tank		0.4			2.8				
6	Mineral Water		0.8	2.2	1.0		0.5	0.3	0.2	1.5
7	Others (Railways supply, broken pipes, drum, mill compounds, etc.)		1.1	0.2	0.6			5.3	2.5	

community tap water is lower than the 54th Round figures of NSSO mainly because it covered all the urban areas in the country, while the present study covers only 7 major cities, having relatively developed water-infrastructures. The Table 13 also shows that some households are not using water from only one source but from a multiple sources. Given the fact that except a few, almost all the households in large municipal cities (under the present focus) being dependent on municipal-source of water supply, it is imperative that these cities augment their water supplies for growing needs of their residents.

It is interesting to note that although in all the seven cities, except in Kanpur, more than 90 per cent of the households have access to municipal tap water, a sizeable proportion of them are also dependent on other sources, indicating the lack of reliability, regularity, and sufficiency of water supply through the municipal taps. Mumbai seems to be the only exception among the seven cities, as in this city only about 4.4 per cent of the total households, besides using municipal tap water, also utilize other sources of water like tube wells (3.6% of the total households), well/open well (0.6%), mineral water (1%), and other sources (0.6%). In contrast, in other cities more than one-fourth of the population also uses water from other sources. In Delhi, 24.5 per cent of the households use tube wells, 2 per cent tankers, 0.2 per cent wells/open well, 2.2 per cent mineral water, and 0.2 per cent other sources of water. Surprisingly in cities like Kanpur, tubewell serves almost 80 per cent of the households. In Ahmedabad and Hyderabad, the proportion of households utilizing the tubewell water or supplementing their needs is 38.5 per cent and 48.5 per cent, respectively. Thus, the second largest source of water for a majority of households in major cities in the country is the ground water. The rapid increase of population in these cities, accompanied by inelastic municipal water supply, makes people more dependent on groundwater leading to its rapid decline. Delhi, Hyderabad, Kanpur are suitable examples in this regard (see also Soni, 2003).

About 10 per cent of the households in Hyderabad are dependent on tanker water supply which is quite costly. Out of these households dependent on water tankers, 46 per cent are dependent on private tankers, indicating that private vendors have their hey-day in the city. About 2 per cent of the households in Delhi and 1.1 per cent in Madurai are also using tanker water supply. Mineral water, particularly for drinking purpose, is also making an inroad in urban households.

This is supported by the fact that 2.2 per cent of the households in Delhi, 1.0 per cent in Mumbai and 1.5 per cent in Madurai use mineral or packaged water. Kolkata seems to be an exception as none of the surveyed households in the city has reported using mineral water for drinking purpose. It is interesting to see that although about 10 per cent of the households in Kolkata say that water is 'not so safe' and 'not safe at all', and further about 26 per cent express their inability to say anything about the quality, of water supplied by municipal corporation, still none of these households reported using mineral water for drinking (Table 14). Ideological orientation of the population, largely supporting communist government and having reservations about multinationals and private companies, may be playing a big role in this regard. Table 14 shows that only about 40 per cent of the households in Kolkata view municipal water 'quite safe' or 'very safe'. In Madurai, Mumbai and Delhi, about 74 per cent, 65 per cent and 52 per cent of the households, respectively, view municipal water supply 'quite' and 'very safe', but in the case of Kanpur, only about 11 per cent say so. Thus, one finds a sizeable chunk of population not considering the municipal water as 'safe' or 'very safe' for drinking.

Table 14

Perception of Households about the Safety of Water from Municipal Taps

(%)

Safety Level	*Total*	*Delhi*	*Mumbai*	*Kolkata*	*Hyderabad*	*Kanpur*	*Ahmedabad*	*Madurai*
Not at all safe	2.9	2.96	.20	3.25	2.76	9.90	1.11	1.89
Not so safe	5.0	6.71	2.40	6.50	3.02	12.54	2.49	1.89
Somewhat safe	21.5	30.97	28.40	24.25	12.06	20.46	7.20	21.13
Quite safe	28.6	30.97	31.80	15.50	22.36	10.23	43.49	47.55
Very safe	25.6	20.51	33.60	24.25	57.04	.99	9.14	26.04
Cannot say	16.8	7.89	3.60	26.25	2.76	45.87	36.57	1.51
Total	100.0	100.00	100.00	100.00	100.00	100.00	100.00	100.00

Contamination of supplied water in the pipe lines has been a common problem, particularly in slums in Indian cities. The data from Municipal Corporation of Greater Mumbai (MCGM) showed that 8-10 per cent of the samples taken at consumer taps in 1997 and 1998 were contaminated by coliform bacteria and unfit for consumption (MCGM, 1999). However, our survey data reveals that about 46 per cent of the households in slums 'feel' that municipal tap water supply is somewhat safer. In comparison to Mumbai, about 37 per cent in Delhi, and only 10 per cent in Kolkata considered that piped water was somewhat 'safe'. As opposed to Mumbai, where none of the households living in slum areas stated that the municipal tap water was unsafe, 2 per cent slum households in Delhi and 2.7 per cent households in Kolkata considered the water 'not safe at all'. In Kolkata, about 49 per cent of the slum households declined to say anything about the safety of the municipal water, while in Delhi and Mumbai, the percentages of slum households expressing the similar views were 26 per cent and 8 per cent, respectively. A study by Karn et al. (2003) in some selected slums in Mumbai shows that the percentage of households boiling water before consumption varies from slum to slum: highest being 14 per cent, and the lowest 5 per cent. However, our survey data reveals that 11.6 per cent of the slum households boil water before consumption, and 80.4 per cent filter it by cloth, and remaining 7.1 per cent do not use any purification method before consumption. In Delhi's slums, 9.2 per cent of the households filter the water before its consumption, 6.1 per cent boil it, and 84.7 per cent use no purification methods. In Kolkata, 79.7 per cent of the slum households use no purification method. In Kanpur, none of the households in slum areas reported using any water purification methods.

Duration of Water Supply and Distance from Source of Water

Cleanliness of municipal water in Indian cities is not the only issue, the supply is also very erratic and for a very limited duration. A basic need and service, like water on tap for 24 hours a day has been unheard of for decades in most Indian towns (ADB, 1993). As the supply is highly erratic and for a very limited duration, the households and housing societies store water in their tanks and drums. In the seven cities, about 18 per cent of the households, who reported using municipal tap water supply, stated that the water supply on tap was available for 24 hours, while about 25 per cent and 27 per cent claimed that it was available for a few hours twice a day, and once in a day, respectively (Table 15). About 21 per cent of the households in these cities report that water on tap comes for a few hours in two days. Among the seven cities, only in the case of Ahmedabad and Kolkata, about a half of the total

Table 15

Availability of Tap Water to Households in Indian Cities

(%)

	Total	*Delhi*	*Mumbai*	*Kolkata*	*Hyderabad*	*Kanpur*	*Ahmedabad*	*Madurai*
24 hours/round the clock	17.9	13.6	5.0	47.3	0.3	5.6	50.1	2.6
For a few hours/ once in a day	27.0	15.4	84.4	2.8	7.0	7.3	38.5	14.7
For a few hours/ twice a day	24.9	64.3	7.4	39.3		44.2	6.4	1.5
Not predictable	1.8	3.7	2.8	0.5	2.0	1.7		
Once in two days	20.7		0.2		88.9		2.2	77.0
Once in four days	0.0							0.4
Weekly twice half	0.1							1.5
Once week two times	0.0							0.4
Once in a day	0.1							1.5
Cannot say	7.3	3.0	0.2	10.3	1.8	41.3	2.8	0.4
Total	100.0	100.0	100.0	100.0	100.0	100.0	100.0	100.0

households report that the water on tap is available for 24 hours. But in case of Mumbai, Delhi, Kanpur and Madurai, only 5 per cent, 14 per cent, 6 per cent and 3 per cent, respectively, of the total households report so. In Hyderabad, hardly any household reports the supply of water for 24 hours. More than 80 per cent of the total households in Mumbai get water on tap for a few hours once in a day, while about two-thirds of the Delhi residents report that they get the supply twice a day for a few hours. Hyderabad, the cyber city, has dubious distinction among the cities as about 90 per cent of its households state that the water on tap is supplied only for a few hours once in two days. In Kanpur, the situation is more pathetic, where more than two-fifths of the households inform that the supply of municipal tap water is not predictable at all.

Only about 68 per cent of the total population in the seven cities under focus have source of water supply within their premises (Table 16). The 54th Round of NSSO data also reveals that a total 65.7 per cent of the households in urban India have source of water supply within their dwelling or premises (Bajpai and Bhandari, 2001). There is wide variation among the cities in terms of location of the source of water. Whereas about 62 per cent and 50 per cent of the households have the source of water outside the dwelling and premises in Kanpur and Kolkata, respectively, in Mumbai it is only about 13 per cent of the households who have the source of water outside their premises. In Ahmedabad and Delhi, about one-fifth and one-fourth, respectively, of the total households collect water from outside their house premises.

Table 16

No. of Trips by Households for Collection of Water from Outside the House Premises

(% to total)

No. of Trips	*All Citiels*	*Delhi*	*Mumbai*	*Kolkata*	*Hydera-bad*	*Kanpur*	*Ahmeda-bad*	*Madu-rai*
1	9.6	6.3	2.4	23.5	8.5	14.5	5.5	10.2
2	9.3	8.1	4.0	17.3	1.8	25.1	5.8	7.2
3	3.5	3.4	1.4	3.0	3.8	8.6	2.8	3.8
4	2.3	3.4	0.8	1.5	0.8	6.9	1.1	3.0
5	4.1	3.2	1.8	2.5	5.3	3.0	2.8	14.0
6	1.2	1.0	0.4	1.5	2.0	2.0	0.3	1.9
7	0.6	0.4	1.2	0.5	0.5	0.3	0.8	0.4
8	1.0	0.6	0.8	0.5	1.3	1.3	1.1	2.3
9	0.04	0.0	0.0	0.0	0.0	0.3	0.0	0.0
Total (%) of Households	31.7	26.4	12.8	50.3	24.0	62.0	20.2	42.8

Table 17

Time Taken for a Trip to Collect Water from Outside the Premises

Time	*Total*	*Delhi*	*Mum-bai*	*Kolkata*	*Hydera-bad*	*Kanpur*	*Ahmeda-bad*	*Madurai*
1 to 5 minutes	26.0	18.1	8.6	45.8	19.8	59.4	14.1	32.1
6 to 10 minutes	3.6	5.5	1.0	1.5	3.0	2.6	5.3	7.9
11 to 15 minutes	0.7	1.6	0.2	0.3	1.2			1.5
16 to 20 minutes	0.3	0.4		0.5			0.8	
21 to 30 minutes	0.9	0.6	1.8	2.3				1.3
Above 30 minutes	0.2		1.2					
Total	31.7	26.4	12.8	50.3	24.0	62.0	20.2	42.8

Table 16 shows that among the households (together in the seven cities) which have source of water outside their premises, about two-thirds make up to two trips to collect water sufficient for consumption for the day. City-wise analysis indicates that about a half of the households in Delhi, Mumbai, Hyderabad, Ahmedabad, and Madurai make three and even more trips to bring sufficient water for their daily consumption. The source of water for a majority of households in these cities is located nearby the residence, as it takes less than five minutes for more than 80 per cent of the households to make a trip for water collection (Table 17). However, at the city-level there are again wide variations in this regard. For example, about 91 per cent of those households who have source of water outside the premises in Kolkata, it takes only less than five minutes for them to make a trip to fetch water, while in Delhi and Mumbai about 31 per cent of such households take more than five minutes to fetch the water.

Storage of Water by Households

The shortage of water, even for daily routine consumption, leads to storage of it by a majority of households. In fact, it is through storing that they can make sure the availability of water for daily consumption. Table 18 shows that the storing upto 500 litres of water by households in these cities has become a norm, and a majority of these households

Table 18

Storage of Water by Households in Different Cities

(%)

Litres	*All 7 Cities*	*Delhi*	*Mumbai*	*Kolkata*	*Hyderabad*	*Kanpur*	*Ahmedabad*	*Madurai*
Up to 50	18.8	16.8	6.8	35.1	9.3	42.6	15.7	2.6
51 to 100	18.9	19.9	9.6	36.1	14.3	25.1	16.0	14.0
101 to 200	18.6	13.8	18.4	15.0	34.2	14.2	14.2	23.0
201 to 500	32.2	36.9	52.6	12.5	35.4	13.5	22.7	46.4
501 to 1000	9.0	12.4	12.6	0.8	4.5	4.3	16.3	12.1
1001 to 2000	1.5	0.2			0.8	0.3	9.7	1.1
2001 to 5000	0.9			0.5	1.0		5.1	0.8
Above 5000	0.1				0.5		0.3	
Total	100.0	100.0	100.0	100.0	100.0	100.0	100.0	100.0

store water between 100 to 500 litres. There is also a wide variation among the cities in terms of storage of water. For example, in Kolkata and Kanpur, about 70 per cent of the households store water up to 100 litres for their daily consumption, while in Mumbai, Hyderabad, and Madurai, about 70 per cent of the households store between 100 to 500 litres per day. In Delhi, about a half of the households store water between 100 to 500 litres.

Table 19

Asset Class-wise Distribution of Storage of Water by Households

(%)

Cities	*Asset Class*	*Storage in litres*								*Total*
		Up to 50	*51 to 100*	*101 to 200*	*201 to 500*	*501 to 1000*	*1001 to 2000*	*2001 to 5000*	*Above 5000*	
All 7 Cities	Very Poor	36.1	26.9	18.5	16.8	1.7				100.0
	Poor	26.9	24.4	20.1	24.6	3.3	0.5	0.3		100.0
	Lower	19.2	20.6	20.6	31.2	7.4	0.9	0.2		100.0
	Middle	11.8	15.4	17.9	39.7	11.8	1.4	1.7	0.2	100.0
	Upper	7.4	11.1	10.6	36.1	22.7	7.9	3.7	0.5	100.0
Delhi	Very Poor	20.0	10.0	30.0	40.0					100.0
	Poor	28.1	40.6	12.5	12.5	6.3				100.0
	Lower	25.1	26.8	16.4	26.8	4.9				100.0
	Middle	12.6	17.3	13.6	42.5	14.0				100.0
	Upper	1.5	1.5	5.9	57.4	32.4	1.5			100.0
Mumbai	Very Poor	10.5	26.3	15.8	42.1	5.3				100.0
	Poor	5.3	12.3	21.1	57.9	3.5				100.0
	Lower	7.7	9.8	21.3	50.3	10.8				100.0
	Middle	4.1	6.6	13.2	56.2	19.8				100.0
	Upper	11.8			58.8	29.4				100.0
Kolkata	Very Poor	42.1	47.4	10.5						100.0
	Poor	49.2	41.5	7.7	1.5					100.0
	Lower	34.8	36.6	17.4	9.8	1.3				100.0
	Middle	26.0	30.1	16.4	24.7			2.7		100.0
	Upper	16.7	22.2	11.1	50.0					100.0

Contd...

Contd...

Hyderabad	Very Poor		25.0	62.5	12.5					100.0
	Poor	12.1	15.5	41.4	25.9	3.4	1.7			100.0
	Lower	10.6	13.2	34.4	39.7	2.0				100.0
	Middle	6.7	12.6	32.6	42.2	3.7		1.5	0.7	100.0
	Upper	10.9	19.6	23.9	17.4	17.4	4.3	4.3	2.2	100.0
Kanpur	Very Poor	83.3	10.0	6.7						100.0
	Poor	63.8	29.0	4.3	1.4	1.4				100.0
	Lower	44.4	33.3	14.4	6.7	1.1				100.0
	Middle	19.6	21.6	21.6	31.4	4.9	1.0			100.0
	Upper		8.3	25.0	16.7	50.0				100.0
Ahmedabad	Very Poor	31.3	43.8	18.8	6.3					100.0
	Poor	18.0	34.0	26.0	12.0	6.0	2.0	2.0		100.0
	Lower	16.5	19.1	18.3	21.7	17.4	6.1	0.9		100.0
	Middle	14.5	5.5	8.2	31.8	21.8	9.1	8.2	0.9	100.0
	Upper	7.5	2.5	2.5	20.0	17.5	35.0	15.0		100.0
Madurai	Very Poor	5.9	29.4	23.5	35.3	5.9				100.0
	Poor	3.2	4.8	28.6	58.7	4.8				100.0
	Lower	1.8	7.0	21.1	50.0	16.7	2.6	0.9		100.0
	Middle		23.2	23.2	37.5	14.3		1.8		100.0
	Upper	13.3	53.3	13.3	13.3	6.7				100.0

As expected, the water is not stored equally by each social class but differs across them. This fact is very well illustrated by the asset-class and water storage class-wise data for the cities, presented in Table 19. The distribution of proportions for almost all the cities shows that the storage of water increases with asset-class of the households. In other words, the rich households store more water than the poor households. At the aggregate level (for all the cities together), where 63 per cent of poor households store 100 litres and below, about 35 per cent of the upper class store 500 an above litres of water. At the aggregate level, the percentage of households in the very poor class, storing above 500 litres of water in all the classes, is almost negligible. The similar pattern of distribution of water storage is seen in almost all the cities, except in Kolkata where the percentage of households storing water above 500 litres is almost negligible. As observed earlier, about a half of the total households in Kolkata have 24 hours municipal tap water supply, and therefore, no need of storing water beyond five hundred litres in the house. In cities like Kanpur, Delhi, Ahmedabad and Mumbai, a sizeable proportion of even upper and middle class households store water 500 litres and above.

Perception about Wastage of Water

It was attempted to assess the perception of households about the activities where wastage of water, in their view, is done. Data presented in Table 20 shows that almost similar kinds of response patterns emerge from each class of people (based on the asset score). About 40-50 per cent of the households in these seven cities under study feel that most water wasting activity is the washing clothes, while 15-20 per cent of the households call bathing as most water wasting activity. Washing utensils and dishes comes next and is followed by use in toilets. There is a significant variation between the poor and rich class about the perception of wastage of water in toilets. Whereas only about 7 per cent of the very poor households feel that they waste some water in toilets, about 11.3 per cent from the upper class households perceive it so. As these four activities mentioned above are the most water consuming activities, a significant achievement in water saving can be made if some technological breakthrough is achieved for saving the water in these activities. Moreover, the household perception is an indication that given the appropriateness, affordability and accessibility to new technology for saving water in these activities, households

Table 20

Perception of Various Categories of Households about the Most Water-Wasting Activities

(%)

Activities	*Very poor*	*Poor*	*Lower*	*Middle*	*Upper*
Washing clothes	47.9	49.1	43.4	40.5	43.0
Bathing	17.4	14.0	15.8	16.4	19.5
Washing dishes/utensils	10.7	10.7	12.7	11.1	11.8
Use in toilets	6.6	10.7	12.4	14.9	11.3
Cleaning house	4.1	8.5	8.4	9.3	7.2
Cleaning vehicles	0.8	0.2	0.3	2.2	2.7
Others	12.4	6.7	7.0	5.6	4.5
Total	100.0	100.0	100.0	100.0	100.0

would be willing to adopt them. The augmentation of water supply alone is not going to solve water problem in our cities; there is also a need for research and development to minimize the wastage and use of water in various activities. Adopting low sud detergent to wash clothes, low capacity tank with higher elevation for flushing toilets, washing utensils immediately after the use, and stopping showers and using bucket of water for bathing can go a long way in saving this precious natural resource.

Table 21

Perception of Households about the Most Water-Wasting Activities

(%)

Activities	*All 7 Cities*	*Delhi*	*Mumbai*	*Kolkata*	*Hyderabad*	*Kanpur*	*Ahmedabad*	*Madurai*
Washing clothes	43.6	33.7	39.2	38.5	41.2	40.3	58.2	65.7
Bathing	16.1	18.9	9.8	27.5	6.0	27.1	14.7	9.4
Use in toilets	12.5	25.4	6.6	5.0	11.8	8.3	14.7	13.6
Washing dishes/utensils	11.8	9.3	28.8	13.8	10.3	5.9	2.5	3.0
Cleaning house	8.4	5.3	5.4	12.0	18.1	10.2	4.7	2.6
Cleaning vehicles	1.1	0.4		0.3	3.5	2.3	1.1	0.8
Watering the garden/lawn	0.5	0.4		0.3	0.3		2.5	
Others	6.1	6.5	10.2	2.8	8.8	5.9	1.7	4.9
Total	100.0	100.0	100.0	100.0	100.0	100.0	100.0	100.0

Table 21 shows that there exists a significant city-wise variation in perception of households about the most water-wasting activities at their household level. Where in Delhi about one-third of the households feel that washing cloths is most the water-wasting activity, in Madurai as many as two-thirds of the households say so. In Kolkata and Kanpur about 27 per cent of the households state that the most water-wasting activity is the bathing; while in Hyderabad only 6 per cent and in Mumbai and Madurai 9-10 per cent of households feel so. About one-fourth of the Delhiites feel that they waste most water in toilets, but in Mumbai and Kolkata the percentages of households perceiving so are less than 7 per cent. In Mumbai and Kolkata, 28.8 per cent and 13.8 per cent, respectively, of the total households feel that the most water washing activity is the washing of dishes and utensils. In the case of Kolkata, Hyderabad and Kanpur, more than 10 per cent of the households think that most water activity at their household level is the cleaning of house (floors). In sum, although there are significant variations about the perception of wastage of water at the household level across cities, the four activities, washing clothes, bathing, use in toilets, washing dishes and utensils, remain universally perceived activities where much of the water wastage takes place.

Prevention of Wastage and Conservation of Water

Water shortage in urban India for all sections of people is a reality. The limited supply or availability has led to a significant proportion of households, in almost all the cities, putting conscious efforts to stop wastage of water. Table 22 shows that more than 85 per cent of the households, irrespective of their status, attempt to stop wastage of water. However, at the aggregate level, all cities, SEC-B has more proportion of households striving to save water in their day-to-day activities.

Table 22

Households Striving to Prevent Wastage of Water

(%)

SEC-Class	*All 7 cities*	*Delhi*	*Mumbai*	*Kol-kata*	*Hydera-bad*	*Kanpur*	*Ahmeda-bad*	*Madu-rai*
SEC-A	90.8	94.0	95.5	97.3	83.8	96.2	79.6	98.0
SEC-B	92.5	97.2	93.3	95.5	84.4	97.6	84.5	93.9
SEC-C	88.2	92.5	88.4	96.2	81.7	97.3	68.6	87.3
SEC-D	87.4	92.8	90.2	93.2	85.1	93.2	64.9	89.0
SEC-E	85.2	89.5	91.5	97.4	83.0	94.1	50.0	86.4
Total	89.4	93.3	91.2	96.0	83.7	96.0	72.9	90.9

Asked about the methods households have been adopting to prevent the wastage of water, 40 per cent of the households reported that they close tap properly and stop leakage in the storage tank; 42 per cent informed that they consumed less water; another 21 per cent responded that they were using only that much water as required. Interestingly, 17 per cent households reported that they have been reusing water for gardening and toilets. The reuse of water, particularly for gardening and toilets has a large potential to solve water problem in urban India, and awareness needs to be spread about this, as only 17 per cent of households presently practice this.

Table 23

Method Adopted to Prevent/Wastage of Water by Various Socio-Economic Groups

(%)

Methods	*All Classes*	*SEC-A*	*SEC-B*	*SEC-C*	*SEC-D/E*
Close tap properly and stop leakage in storage tank	40	36	43	43	28
Consume less water	42	46	43	41	27
Using as much water as required	21	25	21	22	18
Reusing water for toilet and gardening	17	15	17	16	19

Note: Multiple responses possible, and therefore column figures may not add up to 100.

Table 23 shows that these are SEC-D and E, who reuse water to a maximum. We have seen that the consumption by these classes is lower than the other classes, and this is largely due to the limited supply and access to water for these classes. As a consequence, the SEC-D and E adopt the method to carry out their daily activities in the limited availability of water by reusing it.

Table 24

Awareness of Households about Rainwater Harvesting in Different SECs

(%)

	All 7 cities	*Delhi*	*Mumbai*	*Kolkata*	*Hyderabad*	*Kanpur*	*Ahmedabad*	*Madurai*
SEC-A	63.8	58.1	59.1	77.0	71.6	55.8	32.3	98.0
SEC-B	45.6	24.3	49.5	70.0	51.0	47.0	12.4	98.0
SEC-C	40.7	10.0	49.3	48.1	48.3	35.6	5.7	70.1
SEC-D	41.2	8.7	50.9	49.2	36.2	18.2		73.3
SEC-E	28.7	5.3	54.9	43.6	21.3	7.8		76.4
Total	46.0	24.5	51.8	62.0	53.0	35.0	12.7	83.2

The awareness about the rainwater harvesting is also spreading fast in urban India (Table 24). However, the information about availability and awareness also shows class variations. In all the seven cities, where about 64 per cent of the households belonging to SEC-A are aware about rainwater harvesting, only about 29 per cent of household in SEC-E are aware about such a method. In Delhi, the information gap or disparity about awareness of rainwater harvesting is the highest (SEC-A 58.1%, and SEC-E 5.3%), followed by Kanpur (55.8% and 7.8%), and Ahmedabad (32.3% and 5.7%). The least disparity in awareness about the rainwater harvesting method is found in the case of Mumbai (59.1% and 54.9%), Madurai (98.0% and 76.4%) and Kolkata (77.0% and 43.6%). Overall, the awareness is highest in Madurai (83.2%), followed by Kolkata (62.0%), Hyderabad (53.0%) and Mumbai (51.8%). In Ahmedabad, among the seven cities, the least proportion of households reported that they were aware about the rainwater harvesting. In fact, in the city, none of the households belonging to SEC-D and SEC-E reported that they knew about the method of rainwater harvesting. Ahmedabad, Kanpur and Delhi face perpetual problem of water shortage and the tragedy is that in these cities a very limited proportion of households are aware about the rainwater harvesting. Awareness leads to adaptation, and to solve water problem, besides increasing water supply, the local and state governments need to put special emphasis on spread of awareness about various methods of water conservation.

Table 25 shows socio-economic class-wise distribution of households practising rainwater harvesting. It is found that about 73 per cent of the total households covered by the present survey in Madurai practise the rainwater harvesting, while together in all the seven cities, only about 10 per cent of the households practise the rainwater harvesting. In fact in Delhi, Kolkata, Hyderabad Kanpur, and Ahmedabad, the share of total households practising rainwater harvesting is negligible (1% or below). In Mumbai, it is 8.8 per cent. In Mumbai, although, rainwater harvesting is practised by all the SECs and the proportion of households carrying the practice declines as we move from SEC-E to SEC-A, but opposite is seen in the case of Madurai. Among others, the success of rainwater harvesting in Madurai can be attributed to NGOs like Dhan Foundation, and also to efforts by the state and local governments. The efforts of Tamil Nadu State government towards spreading awareness and encouraging practice of rainwater harvesting has been commendable. The Madurai Municipal Corporation Act, 1971 was amended in 2003 for making it mandatory for rain water harvesting structures to every building owned by the government and other statutory bodies.

Table 25

SEC-wise Distribution of Households Practising Rainwater Harvesting

(%)

SEC Class	*All 7 cities*	*Delhi*	*Mumbai*	*Kolkata*	*Hydera-bad*	*Kanpur*	*Ahmeda-bad*	*Madurai*
SEC-A	8.8	0.9	1.5		2.0	1.9	1.1	92.1
SEC-B	8.0		6.7					87.3
SEC-C	12.1		4.8	2.5				67.3
SEC-D	16.9		13.4	1.7				59.2
SEC-E	10.6		19.7	2.6	2.1			56.3
Total	9.9	0.2	8.8	1.0	1.0	0.3	0.3	73.0

Households were also asked that if they practised rainwater harvesting, how much water in general they harvested in a year. Table 26 shows that whereas about 98 per cent of the households reported harvesting 500-100 litres in Madurai, below 50 litres is harvested by 85 per cent of the households in Mumbai. In fact, discussions with such households revealed that in rainy season they collected water in drums and utensils and used it mainly for toilet, and washing utensils. In fact, except in Madurai, the rainwater harvesting in none of the cities is of any significant level (see Tables 25 and 26).

Table 26

Amount of Rainwater Harvested by Households Practising Rainwater Harvesting

(%)

	All 7 Cities	*Delhi*	*Mumbai*	*Kol-kata*	*Hydera-bad*	*Kanpur*	*Ahmeda-bad*	*Madu-rai*
Upto 50	14.4	100.0	84.1	100.0		100.0		
51 to 100	3.3		9.1					2.5
101 to 200	1.3		6.8					0.4
501 to 1000	80.9				100.0		100.0	97.1
Total	100.0	100.0	100.0	100.0	100.0	100.0	100.0	100.0

Summary and Conclusions

Although recycled by nature, water is a limited resource. The model of present kind of development and population explosion have been the main causes of decreasing per capita water availability. To consume more is emerging main goal of the societies, more so of the 'privileged', thus putting further pressure on this diminishing natural resource. Indian cities have been appropriating water resources, traditionally meant for 'subsistence' in rural areas. This process stands accelerated due to high degree of migration of the rural people to large cities in search of livelihood. The urban population, though not even one-third of the total Indian population, is quite large in sheer numbers, viz., around 290 millions. This would need a systematic augmentation of water supply to urban areas, without adding pressure to the available water for rural areas.

The focus of the present paper was mainly to understand the levels of water supply, its consumption and quality in seven major cities in the country, viz., Mumbai, Delhi, Kolkata, Hyderabad, Ahmedabad, Kanpur and Madurai. The study was based on a sample survey of over 2,700 households based on a stratified random sampling. The issues probed in the study ranged from water consumption across the cities under focus, across social classes within each city, level of satisfaction with available quality and quality of water, and city-wise variations in quality of supply of water and its availability through different municipal bodies. Suggestions emerged for future conservation and supply of water to meet needs of people more adequately and efficiently. The key observations are presented here:

- It is observed that water consumption in Indian cities (more so in large cities) is far lower than the norms laid down by Bureau of Indian Standards. Not to mention, in comparison to other major cities in the world, the consumption in Indian cities is also far deficient. The lower consumption results mainly because the water supply not keeping pace with population growth and increasing needs of users.
- It is interesting to observe that though a majority of households consume water below the specified norms, they, by and large, show satisfaction with the available supply. This is mainly because they have delimited their aspiration and requirements of water in relation to available supply from the concerned municipalities or water authorities. Accordingly, they make efforts to conserve water and also show the urge of repairing of leaking pipes and storage tanks within and outside the households. It is also stressed by the respondents that concerned government departments

should reduce distribution losses, including the theft of water from main pipelines.

- Some household activities, like washing clothes, bathing, use in toilets, and washing dishes and utensils are the most water consuming activities at domestic household level in the cities under focus. It has also been found that in these cities, a majority of the households perceive the above activities as most water-wasting activities. This shows that given appropriate and affordable technologies to save water in specified activities, the households would be willing to adopt them. In fact, there exists a large scope for reducing water consumption in washing clothes by adopting appropriate soap/detergent and machines; and in toilets by changing and modifying the flushing system. An awareness campaign about the best practices in these activities can play a big role in saving precious water.
- It is found that information availability is highly class-biased in the large cities in India. The rainwater harvesting methods, which has big potential to solve emerging water crises in the cities, are not known to a majority of people, more so the people belonging to the poorer sections. However, it is found that maximum number of households from lower socio-economic class reuse water, and it is practiced mainly due to limited availability of water to this class.
- As expected, the availability and mode of use of water varies across the socio-economic classes within the cities. But, surprisingly, the difference is not very high. On an average, about 20 litres consumption is higher among the higher classes.
- City-wise variations in the supply and quality of water are very much visible. Water supply in cities like Kolkata and Hyderabad is far better, while Kanpur and Delhi perform the worst. Though it is also true that water supply in Hyderabad is a mix of adequacies and inadequacies, where multiple agencies pitch in to meet the needs. This is mainly because the municipal corporation supplies water to majority of households once, in two days. Water tankers and bore-wells compensate the deficiency of municipal water in this city.
- Twenty-four hour water supply on municipal taps is a dream for a majority of households in the large cities in the country. The

survey reveals that only about 18 per cent of the total households in these cities get 24 hours municipal water supply. The erratic and limited duration of supply of water has become a common phenomenon in these cities. This has forced the households, in majority of the cities, to depend on ground water and other sources of water, like the private vendors who supply water through tankers and drums. These, in turn, are leading to depletion of groundwater due to overwithdrawal by the burgeoning population and emerging water markets. In fact, during the summer season, private water vendors have their heydays in these cities.

- The much talked commodification of water and water services is also making dent into the otherwise role of government departments as the key supplier of water in cities. For instance, in cities like Delhi, Mumbai and Kolkata some households have started using packaged water for drinking purposes. It is also true that so far; none of these cities have permanently handed over the task of supply to their citizens (NGOs) or the private sector.
- To conclude the observations, it can be stated that the supply of water in large cities of India is going to be a big challenge in future. The rapid increase in population in these cities, depleting water resources and enhanced consumer needs are going to create a difficult situation. The market-oriented development is creating new needs in sectors like entertainment, industry and tourism, building industry, adopted new technologies pushing up water needs, more supply in shopping malls, and so on. Simultaneously, the alarming rise in pollution levels in surface water bodies and even in under groundwaters is going to add to the situation. Therefore, an urgent need is felt for a comprehensive water policy for cities which is suitable and satisfactory to growing needs of citizens. The prevailing 'ad-hocism' in protecting, enhancing and conserving water needs to be done away with.

References

Asian Development Bank (1993): "Service Level and the Urban Poor: Managing Water Resources to Meet Megacity Needs", *Themes Paper,* Proceedings of Regional Consultations, Manila.

Bajpai, P. and Bhandari, L. (2001): "Ensuring Access to Water in Urban Households", *Economic and Political Weekly*, September 29, pp. 3774-3778.

Bartram, Howard G. (2003): *Domestic Water Quantity: Service Level and Health*, World Health Organization (WHO), Geneva.

Down to Earth (2005): "Fact Sheet: Fresh Water Up for Grab", *Down to Earth,* March 5.

Figureres, C.M. et al. (eds.) (2005): *Rethinking Water Management*, Earthscan Publications Ltd., London.

Government of India (1997): "Ninth Five-Year Plan 1997-2002, Vol. II", Planning Commission, New Delhi.

– (2002): "Tenth Five-Year Plan 2002-2007, Vol. II", Planning Commission, New Delhi.

Karn, S.K., et al. (2003): "Living Environment and Health of Urban Poor: A Study of Mumbai", *Economic and Political Weekly,* August 23, 2003. pp. 3575-3586.

MCGM (1999): "Health Profile 1997 and 1998", Metropolitan Corporation of Greater Mumbai, Mumbai.

Modi, P.N. (1998): *Water Supply Engineering*, Standard Book House, Delhi.

Ramchandraiah, C. (2001): "Drinking Water as a Fundamental Right", *Economic and Political Weekly,* Feb 24, pp. 619-621.

Soni, Vikram (2003): "Water and Carrying Capacity of a City: Delhi", *Economic and Political Weekly*, November 8, pp.4745-4750.

WHO (2003): "The Right to Water", World Health Organization, Geneva.

9 Environmental Health and Sanitation Facilities: Challenges for India

T.V. Sekher

In most developing countries, three most important environmental health problems that affect a large majority of population are contaminated water supply, inadequate sanitation and untreated solid wastes. The Global Water Supply and Sanitation Assessment Report states that at the beginning of 2000, one-sixth of the world's population was without access to improved water supply and two-fifths lacked access to improved sanitation (WHO/UNICEF, 2000). In the absence of proper sanitation, people suffered from high levels of infectious diseases leading to high incidences of morbidity and mortality. This directly affected the ability of a country to maintain an efficient economy and implied great personal suffering among infected individuals and their families (Richards, 2002). More than a third of world population (about 2.4 billion people) lacked access to adequate sanitation facilities and four out of five of these unserved people lived in Asia (Cairncross, 2003). Inadequate sanitation like unsafe disposal of human excreta, open defecation, lack of infrastructure (sewerage, drainage/sullage), and absence of hygiene management constitute a major threat to the health of the people. Thus, improving environmental health is the most cost-effective means of enhancing people's health and welfare. Nearly 1.7 million deaths each year is attributable

This paper was prepared as part of a project "Restructuring Local Environment Management for Better Health Outcomes: Towards a District Level Pilot in Karnataka", undertaken at the Institute for Social and Economic Change (ISEC), Bangalore and sponsored by the World Bank. Thanks are due to Dr. Shashanka Bhide, Dr. Monica Das Gupta, Dr. Gopal K. Kadekodi, Dr. Manju Rani, Mr. Muhammed Sanaulla, Dr. G.V. Nagaraj and Dr. K.P. Krishnan for useful discussions and Md. Nazrul Islam for assistance in the collection and compilation of data.

to inadequate access to water, sanitation and hygiene (WHO, 2002). Improving urban environmental health can prevent up to 44 per cent of the burden of the disease in the cities of the developing world (Listorti and Doumani, 2001). The problem of poor sanitation will become more serious in the years to come as the city population continues to grow rapidly. It is projected that by the year 2030, the urban population will grow to 3.3 billion, over 90 per cent of which will accrue to the cities of developing countries (UNCHS, 1997).

In the past ten years, diarrhoea has killed more children than all those lost to armed conflicts in almost sixty years since the Second World War. A child dies every 15 seconds from diarrhoea, caused largely by poor sanitation and unsafe water (WSSCC, 2000). Despite the efforts and investment many low-income countries continue to suffer from inadequate and unsafe sanitation.

Figures 1 and 2 show the percentage of population, which has not been served with improved sanitation in the world and in Asian countries including India.

Figure 1
Percentage of Population Unserved (without sanitation) in the World

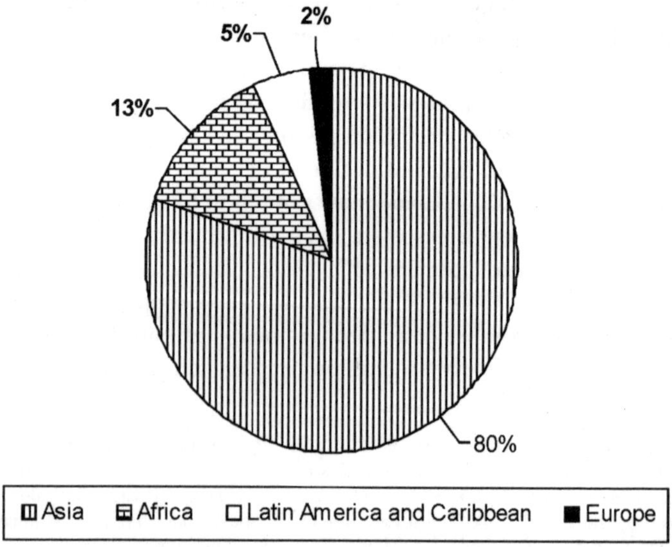

Source: WHO/UNICEF, 2000.

Figure 2

Asia's Unserved Population (without sanitation): Total 1.9 billion

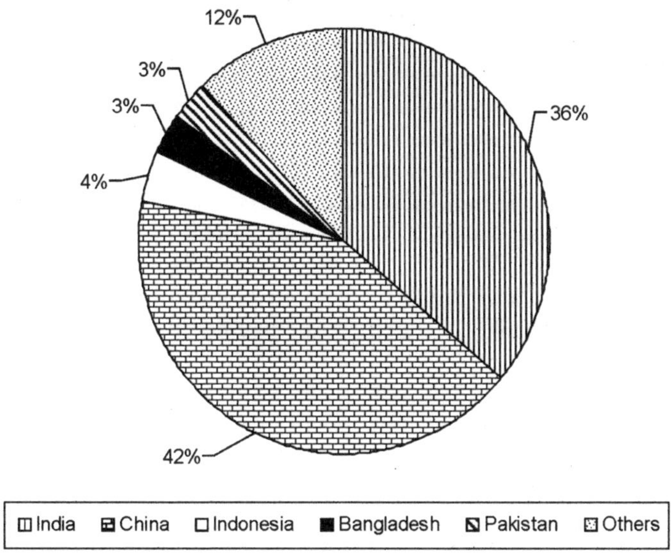

Source: Cairncross, 2003.

The Millennium Development Goals (MDGs) advanced by the international community state that the aim is to halve the number of people without adequate sanitation facilities by 2015. To achieve this target, on an average every day, 350,000 people will have to gain access to improved sanitation between now and 2015 (IRC, 2003). Despite the global commitments, the improvements made by many countries during the last one decade are very poor and the constraints identified are: financial difficulties, institutional problems, inadequate human resources, lack of political commitment, insufficient community involvement, inadequate operation and maintenance, lack of hygiene education, poor water quality, people's attitudes towards sanitation and insufficient information and communication (WHO/ UNICEF, 2000).

This paper attempts to explore the importance of sanitation and its linkage with health status, availability and utilization of sanitation facilities in India, programmes to improve its coverage and experiences in this direction and highlighting the challenges ahead for India by mainly using the available secondary information.

Sanitation in Relation to Health: A Review

The term sanitation has been interpreted differently in different countries and at different times. A WHO Expert Committee defines sanitation as the control of all those factors in man's physical environment that exercise or may exercise a deleterious effect on his physical development, health and survival. According to WSSCC (1998), sanitation means interventions in reducing people's exposure to diseases by providing a clean environment in which to live, with measures to break the cycle of diseases. This usually includes disposing and hygienic management of human and animal excreta, refuse and waste water, control of disease vectors and provision of washing facilities for personal and domestic hygiene. Sanitation involves both behaviours and facilities, which work together to form a hygienic environment.

Poor sanitation and unsafe water have a high health risk but improvement in sanitation has greater impact on vector related diseases than that of improvement of water. Due to lack of proper sanitation, water is contaminated, environment is polluted, vector is increased resulting in major health hazards. Every year more than five million people die from illness linked to excreta disposal and improper hygiene practices (WHO, 2001a). By providing access to adequate sanitation and safe water to all who currently lack would result in 2 million fewer deaths each year from diarrhoea among children under 5 years of age, and 200 million fewer episodes of diarrhoea annually (World Bank, 1992).

Sanitation (sewage, septage, night soil, gray water, and other forms of waste water and their sludges) can contain faecal matter responsible for a broad range of diseases that include diarrhoea like dysentery, gastroenteritis, cholera and intestinal worms like hookworm and ascariasis, tapeworm, threadworm and whipworm, hepatitis, typhoid, polio and range of fevers and blood parasites. In addition, filariasis and scistosomiasis are spread to humans indirectly by mosquitoes and snails respectively which depend on excreta for their life cycles. Many diseases like diarrhoea and some worm infections can be transmitted in several ways. Invariably, faecal infective material enters the human body in or through some medium, viz., drinking water, the hands, soil, food, utensils, and toys. Many possibilities for transmission can be attributed – faecal material may drain into a water source, hands may not have been washed after defecation or before handling food, flies or other insects may transmit infective organisms from faeces to food, etc. Since the organism responsible for most water and sanitation related diseases are present in the faeces or urine of infected persons, the sanitary disposal of excreta is most important. Accesses to sanitary means of excreta disposal and to safe water are universal needs and basic human rights.

Inadequate sanitation, poor hygiene and lack of safe water supply result not only in more sickness and death, but also in higher health costs, lower worker productivity, lower school enrolment and retention rates of girls and perhaps most importantly the denial of the rights of all people to live in dignity (WHO, 2001b; WHO/UNICEF, 2000; Cairncross, 2003; World Bank, 1993). Moraes et al. (2003) found that diarrhoea incidence among children in households with a toilet was half that in households which did not have one. Safe excreta disposal and promotion of hygiene can reduce diarroheal diseases by 36 and 48 per cent respectively (Cairncross, 2003). There was an estimated 25 per cent reduction in the incidence of diarrhoea in children and an average 50 per cent reduction in total population in villages where Water and Sanitation Extension Programme was implemented and services provided effectively (Nanan et al., 2003). Studies also illustrate that there is a strong inverse correlation between access to urban water and sewerage connection, on the one hand, and child mortality, on the other. Increase in the amount of water used and wide coverage of sewerage connection contribute to better hygiene and to the elimination of bacteriological contamination (Shi, 2000). Diarrhoea is somewhat less common among children living in households that boil water or use a water filter for purification of drinking water than among other children (IIPS and ORC Macro, 2001).

The major reasons for the persistence of communicable diseases are non-health determinants such as unsafe drinking water, environmental pollution, poor sanitation and lack of hygiene. Outbreaks of communicable diseases continue to pose a major challenge for public health in India. The incidence of diarrhoea, malaria, cholera, Japanese encephalitis, dengue, and hepatitis occur in large numbers, year after year. In 1998, about 89 lakh diarrhoea incidences were reported in India, including the death of 7,448 persons. Malaria continues to be a major killer in the Indian context, though initial efforts aimed at malaria eradication brought down the cases from estimated 75 million to a record 100, 000 cases in the sixties. Despite medical advancement and the efforts of various anti-malaria programmes, nearly 1,000 malarial deaths were reported in 2001. Table 1 provides the information on incidence of certain communicable diseases in India. The reported mortality due to diseases have been presented in Table 2.

Table 1

Incidence of Diseases in Major States of India

States	*Number of Incidence Reported*						
	Diarrhoea (1998)	*Cholera (2000)*	*Malaria (2001)*	*Japanese Encephalitis (2002)*	*Hepatitis (2000-01)*	*Dengue *(2001)*	*Enteric Fever **(1998)*
Andhra Pradesh	1852642	2	57735	22	27595	1	53252
Assam	596176	-	95142	472	-	-	15330
Bihar	NR	-	4108	8	-	-	NR
Delhi	133089	903	1484	1	4007	322	3219
Gujarat	207027	185	81347	0	3982	69	3617
Haryana	375113	1	1202	59	1086	260	988
Jharkhand	-	-	130784	-	-	-	-
Karnataka	674805	354	197625	152	24571	202	5296
Kerala	550768	146	2289	0	5521	41	9817
Madhya Pradesh	479073	NR	183118	-	6620	NR	39084
Maharashtra	1098750	778	56043	119	40962	54	16004
Orissa	793442	0	454541	-	14011	0	35084
Punjab	196398	14	604	10	1796	49	2827
Rajasthan	211710	13	129233	-	1601	1452	8113
Tamil Nadu	47367	1248	31551	0	1740	816	9067
Uttar Pradesh	564587	0	94524	604	988	11	20929
Uttaranchal	-	-	1196	-	-	-	-
West Bengal	720352	150	145053	0	5831	NR	7414
India	8904597	3807	2085484	1464	153034	3278	318510

Sources: * Annual Report 2002-03, Ministry of Health and Family Welfare and earlier Issues, Government of India.

** Health Information of India, 1997-98, Ministry of Health and Welfare, Government of India and compiled from the website www.indiastat.com

NR: Not Reported.

Table 2

Deaths Due to Diseases in Major States in India

States	*Number of Deaths Reported*						
	Diarrhoea (1998)	*Cholera (2000)*	*Malaria (2001)*	*Japanese Encephalitis (2002)*	*Hepatitis (2000)*	*Dengue (2001)**	*Enteric Fever (1998)***
Andhra Pradesh	674	0	1	3	114	0	36
Assam	121	-	122	150	-	-	0
Bihar	NR	-	0	1	-	-	NR
Delhi	8	0	0	0	162	3	7
Gujarat	50	1	19	0	30	0	7
Haryana	85	0	0	40	6	5	1
Jharkhand	-	-	21	-	-	-	-
Karnataka	366	3	21	15	91	0	7
Kerala	49	11	9	0	8	0	2
Madhya Pradesh	260	NR	81	-	11	NR	47
Maharashtra	556	2	50	16	244	2	32
Orissa	453	0	305	-	107	0	23
Punjab	57	0	0	2	1	0	1
Rajasthan	64	0	36	-	39	35	14
Tamil Nadu	411	1	0	0	4	8	5
Uttar Pradesh	405	0	15	133	24	0	23
Uttaranchal	-	-	0	-	-	-	-
West Bengal	3635	0	191	0	157	NR	17
India	7448	18	1005	361	1038	53	328

Sources: * Annual Report 2002-03, Ministry of Health and Family Welfare and earlier Issues, Government of India.

** Health Information of India, 1997-98, Ministry of Health and Welfare, Government of India and compiled from the website www.indiastat.com

Status of Sanitation in India

Efforts to improve sanitation and hygiene behaviour are not a new development in India. During the British Raj, the frequent outbreaks of fevers and cholera suffered by the British troops had led Sir James Ronald Martin to propose a scheme whereby all medical officers would be required to send reports on sanitary statistics of the districts, stations and cantonments (Ramasubban, 1982). The sanitary perspective and guidelines proposed by Martin was influenced by the growing recognition of clean environment and sanitary practices in England. In 1857, a Royal Commission was appointed to examine "the regulations affecting the sanitary conditions of the army" along with other medical aspects. The large number of deaths of British soldiers due to cholera, diarrhoea, etc., was mainly due to poor sanitary conditions and there was a realization that the main enemy of the British soldier in India was not the Indians but the diseases. To assist all matters relating to improvements of sanitary conditions in barracks, hospitals and stations, three presidency commissions were set up in 1864. After two years, the sanitary commissions were wound up and sanitary commissioners were appointed in each province. The British Government introduced the first Sanitation Bill in India in 1878, which made the construction of toilets compulsory and also proposed the construction of public toilets. But the efforts on sanitary reforms concerning the general population were not enforced fearing that any element of compulsion might offend "the people's customs and religious sensibilities". However, contrary to this belief, the McKenzie committee found that people were willing to submit to any sanitation measures calculated to promote health at pilgrim centres, which were the main centres for the spread of diseases. Lack of governmental intervention resulted in further deterioration in public health and sanitation in India, which compelled the international agencies like the League of Nations and the Rockefeller Foundation to initiate steps to address public health challenges.[1]

The antiquated sanitation systems (bucket latrine) in India used sweepers drawn from the downtrodden communities to empty the buckets, which undermined their social position in the society for centuries. As part of the freedom struggle, Mahatma Gandhi established the Harijan Sevak Sangh for the liberation of scavengers and laid the foundation for a rural sanitation movement in India. The Environmental Hygiene Committee (1948-49) appointed by the government of India carried out an overall assessment and planning of environmental sanitation and recommended a forty-year plan to cover 90 per cent of the population. The suggestions of this committee were never operationalized and no serious attempt was made to improve the sanitation both in rural and urban areas.

It was in 1954 that the rural sanitation programme was introduced in the First Five-Year Plan as part of the health sector. However, the efforts did not succeed until 1980s due to confusion and inconsistency on the sanitation component. Since the beginning of the Sixth Five-Year Plan (1980-85) and the launch of the International Drinking Water Supply and Sanitation Decade in 1980, India has been strengthening its effort to rural water supply and sanitation. Though the provision of water supply and sanitation is the responsibility of the state government, the central government funding constitutes nearly 40 per cent of the total investment in the sector. Water supply has received greater attention compared to sanitation. During the International Water Supply and Sanitation Decade in 1980s, it was aimed to cover 80 per cent of the urban population through formal sanitation facilities and eradicate manual scavenging. Though some progress was made in this direction, the sanitation target set for 1991 under the programme has remained unachieved even now. With the allotment of subsidies and technical assistance for the construction of household toilets during the last decade, there is an overall improvement in the coverage of sanitation amenities.

In the Seventh Five-Year Plan (1985-90), a new programme of sanitation was introduced at the village level for health centres, schools, anganwadis, etc. Individual household latrine constructions were started under a number of government programmes. Tasks like planning, implementing, supervising and coordinating for the Central Rural Sanitation Programme (CRSP) were entrusted to the Ministry of Rural Development in 1986. For the implementation of CRSP, the central and state governments shared the funds. Following the recommendations from the World Bank/UNDP Technology.

Advisory Board, the twin-pit, pour-flush latrine with superstructure became the prescribed toilet technology in India. The Governments provided 80-100 per cent subsidy for the construction of latrines. However, it was soon realized that investment and subsidies alone would not ensure improved coverage of sanitation. In the Eighth Five-Year Plan (1992-97), safe drinking water supply and basic sanitation became vital human needs for the betterment of health. As a major deviation, subsidies were discontinued except for the households living below the poverty line. The sector programme implementation also prescribed better strategies such as user responsibility for the operation and maintenance of facilities, allowing local bodies to charge for sanitation services, etc. The Eighth Five-Year Plan sought to cover 5 per cent of the rural population with sanitation services by the end of the period. It was estimated that less than 4 per cent was covered in 1996.

As a response to this challenge, the Government of India embarked on a programme of Rural Water Supply and Sanitation (RWSS) sector reforms. It is now widely accepted that RWSS services are best managed at the lowest appropriate level. External support agencies and governments all over the world are relying upon community-level organizations in planning, construction and maintenance of sanitation projects. The reform programme of the Government of India makes a paradigm shift from a supply-driven to a demand-driven approach that focuses on decentralized delivery through local elected bodies for the improvement of sanitation. The 73rd Amendment to the Indian Constitution has put enormous power in the hands of Panchayati Raj Institutions (PRIs) and its lowest tier, the Gram Panchayat (GP) has been mandated for the service delivery of RWSS. In many villages, Village Water Supply and Sanitation Committees (VWSCs) coexist with GPs, sometimes as sub-committees and at times independently. VWSCs are user groups of a particular RWSS service and they are meant to maximize user involvement in all phases of water and sanitation projects. The primary aim of the Sector Reform is to generate more demand-responsive and participatory approaches by empowering local governments and community groups to fully manage their own services without any outside support. It is believed that sustainability of the systems can be improved if technical, financial and institutional decisions are made at the appropriate grassroots level.

The Central Rural Sanitation Programme (CRSP) was restructured in 1999 and the Total Sanitation Campaign (TSC) was introduced. TSC aims at providing adequate sanitation facilities to the rural poor, generating awareness about health and hygiene, eradicating manual scavenging by converting all existing dry latrines into low cost sanitary latrines. The TSC is being initiated under the Sector Reform process to promote greater user involvement, lower subsidies, to facilitate NGOs role, and promotion of technologies.

The Ninth Five-Year Plan (1997-2002) emphasized the need for undertaking all possible measures for rapid expansion and improvement of sanitation facilities in urban and rural areas. It was estimated that 20 per cent of the households would be provided with sanitary facilities through the CRSP, through an appropriate mix of central and state investments, institutional finance, strengthening of operation and maintenance system and more importantly, by the involvement of communities at various stages of planning and implementation.

Improvement of sanitation and hygiene promotion becomes one of the biggest development challenges in India. More than 80 per cent of India's rural population has no access to safe and hygienic sanitation facilities. An estimated 650 million Indians defecate in the open and 200,000 tonnes of faecal matter are deposited in the open environment every day.

The Government of India has invested a large amount of about Rs 32,000 crores (6.8 billion) in the Rural Water Supply and Sanitation (RWSS) sector since 1954 but the results and achievements are not very encouraging. A recent review of the sector by the World Bank (1999) has estimated the total investment required for fully covering all rural populations and restoring functionality (repair or rehabilitation) of distressed schemes at about Rs 170-200 billion.

Given this background, it is important to examine the availability and coverage of basic sanitation facilities in India and major states. Considering access to toilet facilities as a better indicator of sanitation at the household level, Table 3 provides a comparative situation of availability of toilets for households in 1991 and 2001 for major states. At the national level, 76 per cent of the households were not having toilet facilities in 1991. This figure came down marginally to 64 per cent in 2001. There exists considerable disparities between rural and urban areas. Though 74 per cent of the urban households have toilets attached to their dwellings, it is only 22 per cent in rural areas. The efforts of the government under various programmes have resulted in a considerable increase in toilet facilities in rural India (from 9.5% in 1991 to 22% in 2001). In the relatively backward states of Orissa, Chhattisgarh, Jharkhand and Bihar, more than 80 per cent of the total households have no toilet facilities. In the states of Madhya Pradesh and Rajasthan it is between 70 to 80 per cent. Nearly 84 per cent of the households in Kerala have latrines attached to their houses in 2001 (for urban Kerala it is still higher at 92%).

Table 3

Availability of Latrine for Households in India, 1991 and 2001

(percentage of households)

States	1991			2001			No Latrine	
	Total	Rural	Urban	Total	Rural	Urban	1991	2001
Andhra Pradesh	18.40	6.62	54.62	32.99	18.15	78.07	81.60	67.01
Assam	37.43	30.53	86.06	64.64	59.57	94.60	62.57	35.36
Bihar	11.75	4.96	56.54	19.19	13.91	69.69	88.25	80.81
Chhattisgarh	-	-	-	14.20	5.18	52.59	NE	85.80
Delhi	63.38	29.60	66.64	77.96	62.89	79.03	36.62	22.04
Gujarat	30.69	11.16	65.71	44.60	21.65	80.55	69.31	55.40
Haryana	22.45	6.53	64.25	44.50	28.66	80.66	37.55	55.50
Jharkhand	-	-	-	19.67	6.57	66.68	NE	80.33
Karnataka	24.13	6.85	62.52	37.50	17.40	75.23	75.87	62.50
Kerala	51.28	44.07	72.66	84.01	81.33	92.02	48.72	15.99
Madhya Pradesh	15.07	3.64	53.00	23.99	8.94	67.74	84.93	76.01
Maharashtra	29.56	6.64	64.45	35.09	18.21	58.08	70.44	64.91
Orissa	9.81	3.58	49.27	14.89	7.71	59.69	90.19	85.11
Punjab	33.18	15.79	73.23	56.84	40.91	86.52	66.82	43.16
Rajasthan	19.57	6.65	62.27	29.00	14.61	76.11	80.43	71.00
Tamil Nadu	23.13	7.17	57.47	35.16	14.36	64.33	76.87	64.84
Uttar Pradesh	18.02	6.44	66.54	31.43	19.23	80.01	81.98	68.57
Uttaranchal	-	-	-	45.20	31.60	86.88	NE	54.80
West Bengal	31.51	12.31	78.75	43.71	26.93	84.85	68.49	56.29
India	23.70	9.48	63.85	36.41	21.92	73.72	76.30	63.59

Note: No latrine - This category covers houses where no latrine of any kind is available to the household within the house. If the household is using sulabh sauchalay or community latrine then it is construed that no latrine is available to the household within the house. However, in the case of multi-household houses, if they are sharing a common latrine, latrine is considered as available to all the households occupying the house.

Table excludes institutional households for which information was not collected. In 1991, Chhattisgarh was part of the state of Madhya Pradesh, Jharkhand was in Bihar and Uttaranchal was in Uttar Pradesh.

Source: Census of India 2001. Series 1, Tables on Houses, Household Amenities and Assets. Census of India 1991 series 1 Part VII, Tables on Houses and Household Amenities.

Census also provides information regarding the type of latrines used both in rural and urban areas. Basically there are three categories: pit latrine, water closet latrine and other latrine mainly including service latrines. Service latrines are dry types of latrines from where human excreta are removed by scavengers. Out of 36 per cent of the households having toilet facilities at the national level, half of them are having water closet latrine, wherein the faecal matter is removed without the need for scavenging. About 12 per cent of the households have pit latrine. In urban India, 46 per cent of the households used water closet latrine and around 15 per cent pit latrine. The trend is in reverse in rural areas where 10 per cent have pit latrine and only 7 per cent use water closet latrines (Table 4). In Uttar Pradesh, more people continue to use service latrine than pit and water closet latrines. In states such as Delhi, Haryana, Punjab and Uttaranchal, the usage of service latrine is quite high even now.

Apart from census, the National Sample Survey Organization (NSSO) also provides information regarding the usage of latrines by households. Among those who are having access to latrines, 74 per cent of the rural and 62 per cent of the urban households use them exclusively and 29 per cent of the urban households have to share the latrines with a set of households. The proportion of households having sole access to the latrine used by them is steadily increasing over the years both in rural and urban areas. In 1973-74, the figure was only 4 per cent in rural and 24 per cent in urban areas. In 1998, it has improved to 13 and 46 per cent respectively. NSSO 54th Round (1998) also indicates that about 7 per cent of the urban households use community toilets (Table 5).

It is important to find out the distance the household members have to traverse to reach the latrine used by them. As per the NSSO estimates, more than 85 per cent of the households using latrine reported that such latrines were located within their premises itself. For 8 per cent of the households, both in rural and urban areas the distance to toilet was less than 0.5 km. However, nearly 1 per cent of rural and 0.5 per cent of urban households reported that they had to travel beyond 0.5 km. to reach the latrine used by them.

Drainage and waste management are interrelated aspects of sanitation. Proper drainage and waste management are essential for keeping the environment clean. Drainage refers to a channel for carrying waste water away from the premises of a building or houses.

Table 4

Types of Latrines at Household Level, 2001

(percentage of households)

States	*Total*				*Rural*				*Urban*			
	Pit latrine	*Water closet*	*Other latrine*	*Total provision*	*Pit latrine*	*Water closet*	*Other latrine*	*Total provision*	*Pit latrine*	*Water closet*	*Other latrine*	*Total provision*
Andhra Pradesh	8.55	18.12	6.32	32.99	6.39	8.62	3.14	18.15	15.10	46.97	16.00	78.07
Assam	43.94	15.90	4.81	64.64	46.91	8.61	4.04	59.57	26.39	58.88	9.33	94.60
Bihar	6.48	7.87	4.84	19.19	5.97	4.16	3.79	13.91	11.37	43.45	14.88	69.69
Chhattisgarh	2.42	8.87	2.92	14.20	1.78	1.83	1.57	5.18	5.15	38.82	8.62	52.59
Delhi	16.36	45.47	16.13	77.96	32.90	19.03	10.96	62.89	15.18	47.36	16.49	79.03
Gujarat	8.73	31.09	4.78	44.60	8.09	11.28	2.29	21.65	9.75	62.11	8.69	80.55
Haryana	22.29	10.91	11.30	44.50	20.46	2.11	6.09	28.66	26.48	30.99	23.19	80.66
Jharkhand	3.27	10.73	5.67	19.67	2.12	2.23	2.22	6.57	7.41	41.24	18.03	66.68
Karnataka	13.38	18.64	5.48	37.50	9.48	4.67	3.25	17.40	20.70	44.86	9.67	75.23
Kerala	12.36	65.19	6.46	84.01	12.78	61.99	6.56	81.33	11.11	74.76	6.16	92.02
Madhya Pradesh	5.93	12.47	5.58	23.99	3.88	2.62	2.43	8.94	11.89	41.10	14.74	67.74
Maharashtra	8.89	21.85	4.34	35.09	10.23	5.33	2.66	18.21	7.08	44.37	6.63	58.08
Orissa	3.97	8.79	2.14	14.89	3.09	3.29	1.33	7.71	9.48	43.05	7.15	59.69

Contd.

Contd...

Punjab	24.33	20.40	12.11	56.84	26.36	6.39	8.16	40.91	20.53	46.52	19.48	86.52
Rajasthan	10.47	11.93	6.60	29.00	8.10	3.18	3.33	14.61	18.21	40.58	17.33	76.11
Tamil Nadu	7.30	23.22	4.63	35.16	4.56	7.36	2.44	14.36	11.16	45.47	7.71	64.33
Uttar Pradesh	10.29	7.98	13.16	31.43	8.34	1.95	8.94	19.23	18.07	31.98	29.96	80.01
Uttaranchal	18.71	15.45	11.04	45.20	16.10	7.17	8.33	31.60	26.72	40.82	19.35	86.88
West Bengal	17.53	20.95	5.23	43.71	15.35	6.97	4.61	26.93	22.89	55.20	6.76	84.85
India	11.50	18.02	6.88	36.41	10.30	7.11	4.51	21.92	14.60	46.12	13.00	73.72

Note: Pit latrine - The latrines attached to the pit that is dug into the ground for the reception of night soil are reckoned as pit latrine. Water closet latrine: The sanitary water flush latrines are those latrines that have water closets fitted with flushing cistern. Such latrines that may be connected to a septic tank or an underground sewerage system will also be recorded as water closet latrines. The faecal matter from these types of latrines is removed without the need for scavenging. Other latrine: Service latrines are dry types of latrines from where a human excreta is removed by scavengers. But, in a few areas, due to some conceptual problems at enumerator's/respondent's level during census, some other type of latrines which do not fall under the typical definition of service latrines, were also found to have been included in this type. Therefore, the data for this type are shown as 'Other Latrine'. Thus, this category includes service latrines; latrines serviced by animals such as pigs, etc., and all latrines other than the pit and the water closet types of latrines.

Source: Census of India 2001, Series 1, Tables on Houses, Household Amenities and Assets. Census of India 1991 series 1 Part VII, Tables on Houses and Household Amenities.

Table 5

Usage of Latrines by Households by States, 1998

(percentage of households)

States	Latrine						Latrine Available					
	For Exclusive Use of the Household		Shared by Restricted Set of Households		For Community Use		Within Premises		Less than 0.5 km Distance		Beyond 0.5 km Distance	
	Rural	Urban	Rural	Urban	Rural	Urban	Rural	Urban	Rural	Urban	Rural	Urban
Andhra Pradesh	71.7	62.6	22.6	34.5	1.2	1.6	89.2	90.5	5.5	4.1	0.8	1.6
Assam	88.2	68.6	4.1	23.5	0.9	2.4	89.5	91.4	9.8	7.4	0.1	0
Bihar	60.0	66.4	13.3	28.9	0.5	1.8	72.5	96.7	5.9	1.6	3.3	0
Gujarat	75.0	73.7	23.4	20.2	0	5.7	82.0	89.6	15.8	10.4	0.4	0
Haryana	78.7	76.8	13.1	22.4	2.2	0.8	88.5	98.9	8.5	1.2	0.4	0
Karnataka	81.5	64.1	16.7	32.4	1.4	3.4	89.1	92.6	8.9	7.3	1.8	0
Kerala	95.4	85.8	3.8	14	0.1	0.1	96.9	98.5	2.2	1.0	0.3	0.5
Madhya Pradesh	51.4	69.4	23.9	16.9	0.9	1.8	73.4	86.8	9.2	3.7	0.3	0.1
Maharashtra	46.3	46.4	20.4	28.5	21.8	24.3	62.3	75.3	24.1	23.6	1.1	0.3
Orissa	77.7	71.4	18.2	25.5	0	0	94.7	87.1	0.1	0.1	0.8	12.8
Punjab	85.6	61.5	11.5	37.1	0.4	0.3	93.2	98.1	4.9	1.5	0.1	0.2
Rajasthan	60.7	65.4	18.9	32.2	10.9	0.8	87.7	98.5	6.3	2.1	1.1	1.4
Tamil Nadu	72.2	59.7	16.4	37.3	0.1	2.4	84.3	91.1	4.6	7.0	2.6	0.1
Uttar Pradesh	68.0	65.6	26.1	25.3	1.0	8.0	88.5	97.4	5.5	2.1	2.1	0
West Bengal	69.6	54.2	21.6	41.9	0.3	2.9	79.7	86.1	8.3	7.5	0.5	0.1
India	74.4	61.9	15.5	29.2	2.6	7.4	85.3	89.2	8.1	8.3	0.9	0.5

Note: Information for nearly 5 per cent of the households not reported.

Source: NSSO, 54th Round (Drinking Water, Sanitation and Hygiene in India), NSSO, Department of Statistics, Government of India, 1999.

Table 6

Drainage Facilities in 2001

(percentage of households)

States	*Total*			*Rural*			*Urban*		
	Closed Drainage	*Open Drainage*	*No Drainage*	*Closed Drainage*	*Open Drainage*	*No Drainage*	*Closed Drainage*	*Open Drainage*	*No Drainage*
Andhra Pradesh	13.65	37.90	48.44	6.29	35.15	58.57	36.02	46.27	17.71
Assam	2.18	18.27	79.55	0.88	14.11	85.00	9.84	42.78	47.38
Bihar	4.46	33.59	61.95	2.61	32.25	65.14	22.15	46.46	31.39
Chhattisgarh	4.41	16.63	78.86	1.42	9.89	88.69	17.12	45.85	37.04
Delhi	49.16	40.78	10.06	13.65	60.88	25.47	51.68	39.35	8.96
Gujarat	27.24	11.60	61.16	6.80	6.86	86.35	59.26	19.03	21.71
Haryana	12.86	63.94	23.2	3.16	68.55	28.28	35.01	53.39	11.59
Jharkhand	6.36	23.26	70.37	1.51	16.21	82.29	23.79	48.57	27.64
Karnataka	17.26	33.97	48.77	4.28	31.11	64.61	41.64	39.33	19.03
Kerala	8.04	11.68	80.28	5.77	10.22	84.01	14.85	16.04	69.11
Madhya Pradesh	7.71	26.48	65.85	1.93	17.90	80.16	24.50	51.42	24.07
Maharashtra	22.05	38.75	39.20	5.15	35.99	58.86	45.07	42.51	12.42
Orissa	4.91	15.84	79.26	2.55	12.30	85.15	19.61	37.88	42.51
Punjab	18.46	63.70	17.84	4.24	73.84	21.92	44.97	44.80	10.23
Rajasthan	7.93	28.58	63.49	2.99	20.18	76.83	24.12	56.07	19.85
Tamil Nadu	16.89	28.23	54.89	4.28	23.07	72.64	34.56	35.46	29.98
Uttar Pradesh	9.54	60.95	29.51	5.29	59.70	35.01	26.46	65.91	7.62
Uttaranchal	10.37	37.69	51.94	4.50	30.46	65.05	28.36	59.88	11.76
West Bengal	7.31	23.45	69.24	1.38	14.54	84.09	21.85	45.30	32.86
India	12.46	33.93	53.60	3.91	30.27	65.82	34.50	43.37	22.13

Note: Connectivity to closed drainage: The house is considered as having connectivity to closed drainage if it has water outlet connected to a closed drainage to carry away the waste generated by the households living in it. Connectivity to open drainage: The house is considered as having connectivity to open drainage if it has water outlet connected to an open drainage to carry away the waste water generated by the households living in it. No drainage connectivity: Such of the houses where waste water outlets are not connected to any drainage system to carry away the waste water are included in this category. Table excludes institutional households for which information was not collected.

Source: Census of India 2001, Series 1, Tables on Houses, Household Amenities and Assets and Census of India 1991, Series 1 Part VII, Tables on Houses and Household Amenities.

Drainage not only takes away the waste water but also takes away storm water. In most villages in India sullage water commonly flows over the village roads. This results in bad smells and spreads dirty sullage on the road. During the rainy seasons the situation may become more dangerous for health. Lack of proper drainage leads to increase in mosquito breeding.

According to the 2001 census, 54 per cent of the households in India had no drainage facilities (66% in rural and 22% in urban areas). The proportion of households with no drainage facility (where waste water outlets were not connected to any drainage system to carry away the waste water) was relatively lower in Delhi, Punjab and Haryana. The Census has made two categories of connectivity of waste water outlet – to closed drainage and to open drainage. Accordingly, 4 per cent of the households in rural and 35 per cent in urban areas had connectivity to closed drainage to carry away the waste water generated by the people living in it. The coverage of open drainage facility was 30 per cent in rural and 43 per cent in urban areas (see Table 6).

Disposal of Household Refuse

Waste management is also an important aspect of sanitation and includes collection, handling, transport and disposal of wastes generated. Due to ineffective collection and disposal of waste material, rodents and flies breed resulting in the spread of diseases. The untreated wastes can pollute surface water and groundwater as well as ferment slowly and can produce bio-gas that can result in adverse health and environment consequence (Bartone et al., 1990). Effective waste management can result in the improvement of health and reduction of morbidity and mortality, improvement of water and air quality and economic development.

The proportions of households reporting removal of their household waste by different modes have been estimated by NSSO. The information on this item has been recorded under four categories (see Table 7). Most of the households (both in rural and urban areas) report that their members are responsible for the removal of garbage from their houses. In urban areas, local authorities and private arrangements also play a major role in the removal of garbage. It is equally important to examine where the household refuse is shifted to after being taken away from the house. About two-thirds of the rural and

Table 7

Arrangement for Removal of Garbage from Houses, 1998

States	*Percentage of Households by Removal of Garbage by*							
	Local Authorities		*Private Arrangement among Residents*		*Household Members*		*Other Arrangements*	
	Rural	*Urban*	*Rural*	*Urban*	*Rural*	*Urban*	*Rural*	*Urban*
Andhra Pradesh	2.4	14.5	2.1	8.1	94.4	75.3	1.0	2.2
Assam	-	4.0	1.2	9.3	91.9	78.1	6.7	8.7
Bihar	0.1	2.0	3.2	7.6	87.1	82.6	9.5	7.6
Gujarat	1.2	28.8	0.8	9.1	90.9	60.9	7.1	1.3
Haryana	0.1	8.4	0.8	14.6	90.9	76.9	7.1	0.1
Karnataka	0.3	20.4	1.0	4.8	96.8	70.3	1.9	4.5
Kerala	0.1	2.4	0.6	1.9	97.7	93.4	1.6	2.3
Madhya Pradesh	0.2	5.7	4.8	6.7	91.3	84.9	3.7	2.6
Maharashtra	0.7	6.9	0.5	22.6	98.1	65.2	0.7	5.3
Orissa	0.1	3.0	0.4	0.4	98.4	96.4	1.0	0.2
Punjab	0.6	3.4	2.5	13.6	84.6	78.7	2.2	4.3
Rajasthan	0.9	15.1	2.5	8.8	94.8	74.6	1.8	1.5
Tamil Nadu	1.7	17.9	0.6	3.0	96.0	76.4	1.7	2.7
Uttar Pradesh	0.5	14.4	2.1	14.5	96.3	69.0	1.1	2.1
West Bengal	0	28.7	0.7	8.8	95.5	59.7	3.6	2.8
India	0.7	13.7	1.8	11.9	94.2	71.2	3.2	3.2

Note: Mode of collection of garbage from house: Information on this item was recorded under four categories: (1) by local authorities; (2) by private arrangement among residents; (3) by household members; and (4) other. While the first category is self-explanatory, the second category included cases where a group of residents (including the households) of the area made the arrangement for garbage clearance. The third category included cases where the household made its own arrangement as also cases where the household members dumped the garbage in a common dumping spot away from the house and the local authorities arranged for its removal from the dumping spot. The category other included cases of removal of garbage by charitable bodies. Site where garbage was taken: Such sites were categorized as: (1) bio-gas plant or manure pit, (2) community dumping spot, (3) household's individual dumping spot, and (4) other. While the second category is self-explanatory, the third one included cases where the household dumped its garbage indiscriminately or where a small number of households dumped their garbage in a particular spot. The first category included cases where the garbage was first taken to a community dumping spot and from there to a biogas plant.

Source: NSS, 54th Round (1999).

one-third of the urban households reported that their wastes were taken to individual dumping spots. A major proportion of the urban households (47%) reported the removal of their waste to community dumping spot and another 7 per cent stated that the garbage was removed to a biogas plant or manure pit (Table 8). With respect to the removal of garbage and waste management, the available data illustrate that we have made very little progress even in urban areas.

Table 8

Arrangement for Shifting the Garbage after Removal from the House, 1998

States	*Percentage of Households Reporting Garbage Shifted to*							
	Biogas Plant for Manual Pit		*Community Dumping Spot*		*Households Individual Dumping Spots*		*Others*	
	Rural	*Urban*	*Rural*	*Urban*	*Rural*	*Urban*	*Rural*	*Urban*
Andhra Pradesh	5.7	1.1	8.7	55.9	69.4	29.8	16.1	13.1
Assam	0.2	0.1	0.4	20.4	79.5	55.9	19.7	23.6
Bihar	1.6	1.6	1.3	14.3	55.9	47.9	40.9	35.9
Gujarat	3.7	2.8	12.9	57.2	42.4	22.0	41.0	17.9
Haryana	5.4	5.6	0.9	20.0	1.4	6.7	1.7	8.8
Karnataka	3.2	1.0	4.6	52.9	80.8	26.1	11.5	20.0
Kerala	0.3	2.2	0.2	7.6	76.2	75.2	23.4	15.0
Madhya Pradesh	16.1	1.5	1.0	35.0	67.0	33.7	15.7	29.6
Maharashtra	14.8	1.9	3.9	70.4	61.8	11.5	19.5	16.0
Orissa	0.5	-	1.0	26.3	83.3	61.5	15.0	12.2
Punjab	4.7	0.2	11.1	45.4	43.4	15.2	40.7	39.2
Rajasthan	12.8	0.1	2.9	32.7	65.4	27.7	18.9	39.5
Tamil Nadu	3.8	3.1	7.7	49.1	71.6	32.8	16.8	14.9
Uttar Pradesh	10.6	0.4	1.6	35.0	66.1	32.2	21.7	32.4
West Bengal	3.8	0.3	0.7	49.8	77.6	36.6	17.2	13.1
India	6.9	1.4	3.5	47.2	66.6	29.6	22.8	21.7

Source: NSS, 54th Round (1999).

However, there are efforts by various agencies in selected cities/ towns to address the challenges of waste management. In India, urban solid waste management comes under the purview of local municipal bodies who are responsible for the collection, removal and disposal of garbage from public places and maintenance of dumping grounds. Sometimes, non-governmental and community based organizations assist the municipal authorities in collecting, treating and disposing waste. In countries like India, the recycling activities are marked by the involvement of informal sectors comprising of waste pickers, itinerant waste buyers and middlemen like the junk dealers and wholesalers.

Collection and Treatment of Waste

There is substantial increase in the generation of waste both in absolute terms and in terms of per capita. Most of the developing countries have insufficient collection and improper disposal of waste. In India, about 0.1 million tonnes of municipal solid waste is generated every day (approximately 36 million tonnes annually). Per capita waste generation in major Indian cities ranges from 0.2 kg to 0.6 kg. The difference in per capita waste generation between lower and higher income groups range between 180 to 800 grams per day. According to the Government of India (2000), the waste collection efficiency in Indian cities ranges from 50 to 90 per cent and out of the total municipal waste collected about 94 per cent is dumped on land and 5 per cent is composted. Urban local bodies in India spend about Rs 500 to Rs 1,500 per tonne on solid waste for collection, transportation, treatment and disposal. During the last five decades there has been eight-fold increase in the total waste generated.

Most of the waste in India, according to the CPCB (2000), is composed of vegetative matter (43.95%) and inert material (41.81%), including fine dirt, stones, and ash. Most of the recyclable material – paper, cardboard, plastics, and metal – is frequently recycled either by households or more often by rag pickers from the informal recycling sector. Nearly 100 per cent of the collected waste is disposed of in landfills/open dumps in India. The dump sites are a major health hazard to rag pickers, workers, and nearby residents. Many of the urban areas will reach the capacity at their dump sites in the near future, or are already continuing to dump beyond the planned lifespan.

Sanitation: Innovative Experiences

The sanitation sector throughout the world is continuously developing and testing new strategies to identify the factors that promote sustainable sanitation facilities (see Boxes 1 and 2). A key lesson learnt from global experience over the past decade is that demand-responsive approach can lead to overall project effectiveness and community level organizations can play an important role in motivating and facilitating sanitary practices (Narayan, 1995; Sara and Katz, 1998).

Evidence shows that successful sanitation projects must have a participatory component. It is also important for the community to feel a sense of ownership for the resource; the community as a whole should believe that this is their project and they are responsible for keeping it operational. Village level committees can enhance demand-based approaches by bringing decision-making down to the village level where users can decide the type of technology, location of the facility, user charges and how they should be used.

Many innovative approaches have been experimented in different states of India for the promotion of sanitation and better health outcomes. Sant Gadge Baba Village Sanitation Campaign in Maharashtra, Midnapur experiment in West Bengal, clean city campaign in Surat are some of the well-known sanitation experiments in India (see Boxes 3, 4 and 5). Sant Gadge Baba Campaign is a good example of how local communities can achieve sanitation without financial support/subsidies from the government.

The Midnapur experience shows that partnership between local governments and NGOs can achieve expected results in the promotion of sanitation as well as public health. Similarly, the Kerala community-managed sanitation programme also reveals that local self-governments and community groups can successfully promote environmental sanitation (see Box 6).

Many factors are responsible for the successful implementation, optimum usage and maintenance of sanitation facilities. Without community mobilization, the sanitation programme cannot be successful. People's active involvement in decision-making plays a major role in ensuring success and sustenance of the project. Capacity-building of volunteers and employers is another important factor that determines the success of the projects.

The experience of community managed sanitation programme for poor households and schools in Kerala (see Box 6) also highlights the need for a decentralized and participatory strategy. The better coordination between local government and community is evident in

Box 1
A Successful Approach to Village Sanitation in Bangladesh

An innovative initiative piloted by Water Aid Bangladesh (WAB), in collaboration with a local NGO called Village Education Resource Centre (VERC), funded by DFID-Bangladesh, has resulted in 100 per cent sanitation in five Thanas in Rajshahi District in Bangladesh. VERC ensures change in existing habits and behaviour patterns of the communities by using and maintaining hygienic latrines, washing of hands, keeping food and water covered, using safe water and maintaining a clean environment. Without any financial assistance or incentive for the communities the project became a success. People were encouraged and motivated to have sanitation. The villagers became genuinely interested in improving their own sanitation status. VERC created awareness among villages about open defecation and its dangerous impact. Community was motivated to change their traditional behaviour by a felt need and a fear of diseases. Communities have begun to share their knowledge with others with a genuine sense of urgency. People visited the defecation sites and realized the situation, calculated the availability of open human excreta that made them repulsive and motivated to change the practices. Once the people were mobilized and sensitized towards behavioural change, the community-promoted sanitation ultimately led to improvement in health. People started to move away from open defecation to safe defecation. VERC is responsible for changing the traditional behaviour of members of the community and igniting awareness about the stark reality of the village sanitation by involving entire villagers in the process. VERC developed and strengthened community based institutions consisting of all sections of people including women, religious leaders and teachers. It chalked out the action plan, organized meetings, and sorted out collective response and solutions. This successful approach illustrates that creative awareness and mobilizing people are crucial components of the village sanitation.

Sources: WSP-SA 2002; DFID 2002.

Box 2
Orangi Experience in Pakistan

Orangi is one of the Pakistan's largest squatter settlements with a population of about 800, 000. The people of Orangi come from a variety of backgrounds. Over 200, 000 refugees from East Pakistan (Bangladesh) were settled and other residents came from impoverished rural and urban backgrounds. The monthly household income varied from Rs 1,000 to Rs 4,000 per month with an average income of Rs 2,500. Before the project started, a majority of the people of Orangi used bucket latrines that were emptied by scavengers every fourth or fifth day for a fee of Rs 15 per month. The scavengers frequently emptied the buckets into the unpaved streets, causing health hazard. The street was filled with excreta and waste water, making movement difficult on the street. Some people had laid sewerage pipes from their houses to the nearest creek or nullah in a haphazard way and the system kept excreta out of the lanes.

A sanitation project was started in 1980 as a result of an understanding between Akhtar Hameed Khan, an eminent social scientist, and Aga Hasan Abedi, President of the Bank of Credit and Commerce International, which provided funds. Organizers and the people of the project area preferred an underground sewerage system for the improvement of sanitation and persuaded the Karachi Development Authority (KDA) to provide sewerage facilities as it did in other richer areas of the city. Dr. Khan spent months to approach KDA to provide such services but failed. Khan and Abedi decided to encourage the Orangi community to build its own sewerage system. The first step was to create a community organization. Each lane had 20 to 30 houses whose inhabitants knew and trusted each other.

Organizers met with the residents of the lane, discussed the sewerage plans and encouraged them to form a lane committee and elect their manager. Once the committee had been established the technical committee surveyed the lane, set up benchmarks and prepared plans and cost estimates for the sewerage work. Lane managers then had the task of collecting money for the scheme from the residents. All the community investment in sanitation was collected, managed and spent by the lane organizations without any book keeping account. In many cases old women acted as treasurers and provided money for the lane manager. Total investment by the community was about Rs 10 million. Each household spent about Rs 500 for the shared facility of primary and secondary drains. It is claimed that this community based approach reduced sewerage costs by 80 per cent compared with official costs. The lane organizations were responsible for O&M of the sewerage system and it was their duty to clean and repair manholes and sewers. A major environmental and social change has taken place in Orangi. The lanes which had a sewerage system are now cleaner and healthier.

Sources: Hasan 1988; World Bank 1999; Serageldin 1994; Pasha and McGarry 1989.

Box 3
Sant Gadge Baba Clean Village Sanitation Campaign in Maharashtra

Maharashtra Government is playing the key role in running the Sant Gadge Baba Village Sanitation Campaign (SGBVSC). The idea is to create an atmosphere which motivates people to become the driving force in sanitation efforts, while promoting new habits that can be sustained, thereafter. Gram Panchayats (GPs) play the central role in the SGBVSC. The SGBVSC is not a programme or scheme and is a campaign to educate and motivate the rural communities. SGBVSC is a contest, whereby communities compete against each other in order to gain a reward, to be used for the collective good. Points are earned as a variety of cleanliness principles are implemented, ranging from toilet use to personal hygiene. In the competition, the three highest-scoring villages at each level (Block, District, Region and State) are awarded with a cash prize to be used for the benefit of the entire community, though not necessarily for sanitation purpose.

The main benefit to the villages involves reputation and recognition rather than simply monetary gain and communities stand to lose their recognition if they do not maintain their village in ways that are environmentally sound. Interested villages register to participate in the competition and to implement various specified works that lead to an environmentally clean village. The village through its own resources and labour undertakes all the works. The villages are then evaluated by independent committees on the specified criteria and other aspects like equity, innovations, etc. The villages that score highest marks are awarded 'prizes' at different levels. Those villages which do not 'win' do not receive either cash or subsidy from the Government. In addition to other works, an estimated one lakh household latrines are being built in the state as a result of this programme. A total expenditure of Rs 6.6 crores in the form of cash prizes leverages the creation of an estimated Rs 200-250 crores worth of rural infrastructure, annually, through collective community action.

Sources: DDWS 2002; WSP-SA 2002; ISEC, 2005

Box 4

Sanitation in Midnapur District in West Bengal

Experience in West Bengal draws attention to the importance of the role of local government in collaboration with a local NGO called Ramkrishna Lok Siksha Parishad (RKMLSP) in facilitating sanitation improvement in the district. The programme was supported by the UNICEF. The dedication, interest and coordination of the local government and the NGO were the driving forces to ensure success in sanitation in the district. The coverage of households by sanitary latrines in the district had increased to 60 per cent in 2001 from 5 per cent in 1991. This success could be attributed to effective cooperation in the provision of hygiene and health education, regular monitoring of the activities. WATSAN (Water and Sanitation) Committees have been formed at District, Block and Gram Panchayat levels.

The Zilla Parishad then carefully selects the NGO who will best carry out the field-level projects, monitoring and evaluation, while the Gram Panchayats select the motivators who then establish direct contact with the households. Every week, scheduled meetings are held at Gram Samsad, Gram Panchayat, Block, Sub-Division and District levels in order to maintain close lines of communication, as well as strengthening the capacity at each level. This programme uses IEC as a means of informing and drawing commitment from the community. The programmes are implemented under the supervision and guidance of panchayats. The RKMLSP implements the programme with the help of youth clubs who promote the concept of hygienic sanitation in their areas. Nandigram II Block in the district has achieved the distinction of being the first block in the country to have saturated all rural households with sanitary toilets.

Sources: WSP-SA 2002; Government of India 2003; World Bank 1999; Roy 1996.

Box 5

'My Surat - Clean Surat': A Sanitation Campaign in the City of 'Plague'

Plague hit the city of Surat (Gujarat) in September 1994, known for its filthiest slums swollen with migrant workers. However, in 1997, Surat (population 3.4 million) was ranked by India's heritage trust as the country's 'second cleanest city'. Municipal administrators have been streaming in to learn from the mobilization campaign by the City Commissioner Mr. S. R. Rao. He attributes the success "to the people of Surat, their representatives, the 15, 000 employees of the Surat Municipal Corporation, the press and the judiciary". He encouraged the authorities and leaders and citizen groups to get involved in the making of new Surat. He also demonstrated that his corporation could 'deliver'.

Sanitation has been a key focus in the 'My Surat - Clean Surat' campaign which also targets 18 other action areas. Daily fieldwork is organized by city zones, each of them networked with a central control system through computer and radio links. Everyone, from commissioners to cleaners, is expected to be out on the job each day between 7:30 am and 12:30 pm. A micro plan for sanitation divides the city zones into sectors of 3,500 m^2, each with its own supervisory and task forces. Public latrines and urinals are cleaned each day, while every afternoon another group of cleaners moves out to follow up on the morning's activity. Special ward maps help these teams pinpoint critical locations. Defaulting citizens have to pay administrative charges for cleaning up after them, ranging from Rs 50 to 5000, depending on the mess. At 3:00 each afternoon, Surat's 15 commissioners meet, armed with 9-page computerized reports for a "free and fair discussion and joint decision-making". Sharing of experiences and random cross-checks are especially encouraged.

Over 50 'Pay and Use' toilets for men, women, and children operate through private initiative, and include the participation of the well-known Sulabh Corporation and Akhil Bharatiya Paryavaran Sansthan. Around 1,600 public "pay" toilets were constructed and more than 90 per cent were located inside the slums. They can be used free of charge by women and children, and males over 12 years pay Re 0.50, a very small sum. People demonstrated their willingness and ability to pay and this small payment has kept all of these units operating. The feedback system operating out of each ward office includes deadlines for responding to categorized complaints. This is 48 hours for cleanliness of public toilets and cesspool overflow, and 24 hours for solid waste disposal. City media have been mobilized to keep a close tab on the progress and help educate the public regarding new patterns of behaviour. Eighty per cent of Surat's slums have now been provided with sanitation and other basic facilities.

Public support has made sanitation a political issue for the first time. Surat demonstrates what people and the political system can do together if there is will on both sides. 'The Surat miracle' has been achieved within the constraints of existing administrative and financial procedures. All the money required for this change came from funds available in the normal budget supplemented by funds raised by the citizen's groups brought together by the Commissioner. No state or central funds were divested for sanitation. In other words, the new sanitation effort is sustainable for the long-term. Naturally, one may ask – "If Surat can do it, what excuse does that leave other cities for not following suit?"

Source: WHO (2004).

Box 6
The Community-Managed Sanitation Programme in Kerala

The Socio-Economic Units (SEUs) of the State of Kerala initiated a programme in 1988 with support from the governments of the Netherlands and Denmark to assist the Kerala Water Authority (KWA) to promote community approaches for effective implementation and management of sanitation facilities. The programme emphasized the promotion of latrine constructions, handwashing with soap, construction of drainage at public taps, chlorination of traditional wells and improvement of school sanitation. The goal was to provide poor households with permanent latrines of good quality, in such a way that they appreciated the facility and would use it properly. The programme has given priority to the mobilization and motivation of the users, and promotion and monitoring for good practices that includes flexibility in planning, negotiation with local elected bodies, decentralized and local management, partnership (involving personnel affiliated to all major local institutions – schools, nursery schools, clinics, women's or youth groups), education and capacity-building, financial contributions from local governments and households, reliance on local materials, local masons, and competitive tenders. The local committee undertook a wide range of activities related to education and sanitation. These activities include mapping, site selection, organizing education programmes, home visits, and managing household selection, purchase, transport, etc. The committee monitors the use and cleanliness of the latrines and the indicators for other good hygiene.

The physical achievement of the programme was 35,500 latrines construction by September 1995, of which, 96 per cent of the latrines were being used. These latrines facilitated about 200,000 low-income (below poverty line) people with hygiene education programme. The SEU succeeded to cover a total 60 per cent of poor population in the programme areas. This achievement was more than double compared to the achievement of other programmes which were implemented before the SEU. Each SEU serves an average 20 panchayats (a population between 400,000 to 1,000,000). About 85 to 98 per cent of the latrines in each area are found very clean, more than 80 per cent have water stored within or very nearby the latrine and about 20 to 70 per cent of the families have soap available at the latrine. A total of 274 school health clubs had been formed and was operating under this programme. The school health club is usually composed of all children in the fifth standard (about 10 to 11 years of age). The school health clubs are involved in a wide range of activities such as ensuring the school grounds clean and waste containers (which are used) in each class, monitoring the use of latrines and helping to clean them, monitoring washing of hands and cleaning of containers for food, monitoring the school water point and ensuring correct use, and holding campaigns for special health issues in the school and community. The parent-teacher associations pay 25 to 50 per cent of the cost of latrine facilities in the school. The project considers the school health clubs, a majority of which are very active, to be a good investment in safe hygiene behaviours of the future generation. The success of this experimental project lies in its involvement of people of all political and cultural affiliations. The focus has been on developing a decentralized strategy managed by local governments and voluntary community groups.

Source: IRC International Water and Sanitation Centre. 1996.

the case of Borban Village in Maharashtra where a successful sanitation campaign has resulted in overall improvement (see Box 7).

However, in some cases, it is observed that there is a conflict and lack of cooperation between local government and the committees set up specifically to implement the programmes, like village water supply and sanitation committee (VWSC). But, a better understanding between them will enhance the success of programme implementation, as observed in many places.

When people are able to access toilet at affordable price, they are willing to use it. Sulabh sanitation experience is an example – Sulabh has been constructing latrines with affordable price and comfortable technology for decades. Development of local enterprise related to sanitation services is also essential. When local enterprises are available for supplying and constructing toilets at low prices, the community may make use of it (see Boxes 8 and 9).

Box 7
SGBCV Campaign: Impact Visible at Borban Gram Panchayat

In the Borban village in Ahmednagar district of Maharashtra, there was no drainage system. Drain water used to be flowing into the street causing foul smell and breeding of mosquitoes, and open defecation was common causing water and air pollution. There were only 14 toilets in a village of 726 people. Women used to suffer from stomach/constipation problem. As a result of SGBCV Campaign, drains were constructed; there is now open gutter system of 700 ft long. A 'Jumbo Soakpit' (13′ x 13′ x 13′) is used for draining waste water. Waste water from every household is collected in the gutter with pipes connected to household outlets. About 65 families channel their drain water into their household garden. There is no open defecation; there are 108 household toilets, and one community toilet. Loans were provided for constructing household toilets. 94 toilets were constructed within 25 days under the campaign. The toilets are indeed used.

The village was kept neat and clean and slogans highlighting the importance of cleanliness could be seen on the walls of village buildings. The place, which was earlier used for open defecation, is now converted as playground. The village is called as 'Hagandari Mukt Gram' (Meaning village freed from open defecation).

Undoubtedly, the gram panchayat played a major role in convincing the locals and triggering community actions. The gram sabha and mahila sabha were instrumental in mobilizing public opinion towards total sanitation. In brief, it was people's movement for a healthy and hygienic living.

Source: ISEC (2005)

Concluding Remarks

The discussion so far has dealt with the status of sanitation in India, different programmes, experiences and achievement. The coverage of sanitation facilities in India is far behind many other developing countries. Lack of sanitation and hygiene is the primary cause of most of the infectious diseases, which are rampant resulting in high

Box 8
Sulabh Sanitation Experience in India

Sulabh International Social Service Organization (SISSO), a NGO, founded by Dr. Bindeshwar Pathak has demonstrated the success of low cost, pour flush, water-seal toilet with twin leach pits for on-sit sanitation technology throughout the country. This technology is affordable even by the economically weaker sections of the society and is designed to suit different levels of income groups. SISSO had constructed over 800,000 twin pit flush toilets by 1997. It has made nearly 240 towns scavenging free where about 40,000 scavengers have been relieved of the degrading human practice of manual removal and transport of human excreta. In the towns which have become scavenging free, all dry or bucket latrine have been converted into the twin pit pour-flush toilets. It has built and maintained about 3,200 community toilets which are run on pay-and-use basis. Over 10 million people use the sanitation facilities provided by SISSO daily. A key to the success of SISSO lies in creating public awareness and seeking community participation in implementing and maintaining the infrastructure. It is also involved in research and development activities to promote low cost sanitation technology and the method to reach the people for its successful implementation.

Sources Government of India (2001); Pathak 1997.

morbidity and mortality. Even after many decades of concerted efforts and investment of huge resources, the achievements are limited and challenges are many. The British introduced the first Sanitation Bill in India in 1878 which made the construction of toilets compulsory even in the huts of Calcutta (Kolkata), the then capital of India. In 1993, Sanitation Act made dry latrine and its manual cleaning a punishable offence. But, the shocking reality is that nearly six lakh scavengers continue to lift night soil from over 72 lakh latrines in 258 towns of India (Krishnakumar, 2003). According to the WHO estimates, given the technology presently used and the current rates of construction and population growth, it would take 200 years for every Indian to have access to a toilet. Kenneth J. Cooper (1997) observes: "while cities of an

Box 9
WaterAid India Sanitation in Tiruchillapally District in Tamil Nadu

WaterAid India, with its NGO partners, creates an enabling environment for the successful sanitation programme. Community groups are facilitated to provide fundamental sanitation and hygiene information. Technology is made available at an affordable price to accommodate the growing demand for sanitation facilities. Both production and promotion are decentralized so as to allow for competitive pricing and innovations. Credit and other financial opportunities are also provided for those in need. These are not disbursed as a form of reward but rather as an enabling factor for those who might otherwise not be able to afford a toilet. The individual subsidy that comes from the Government is redirected to a village development fund based on community consensus and decision. The community for common village development activities manages the fund. Banks and local financial institutions recognise sanitation as a priority investment and credit is available to community groups at market interest rate. No household is excluded and both the burden and benefits are shared through cooperation

Source: WSP-SA 2002.

early civilization in the Indus River Valley had sophisticated sewer systems and among the oldest known toilets in the world-brick models that date back 4,500 years, today, over two-thirds of Indians do not have access to toilets".

Many factors can be identified for the ineffective implementation of sanitation programmes in India. Sanitation and hygiene are given very low priority by the governments in India. Even Gram Panchayats seldom consider improving sanitation as a necessity (Sekher, 2003). Low awareness of health hazards, unaffordability of household toilet, lack of technology innovation in designing toilets, inadequate emphasis on Information, Education and Communication (IEC), heavy reliance on subsidy, and above all, scarcity of water are the major reasons for the poor progress in developing sanitary infrastructure. Sanitation was never considered a requirement in most of our rural areas. We need to take up sanitation issue on a priority basis. The Total Sanitation Campaign (TSC), launched by the government of India in 1999, was a new attempt to address the neglected dimension of 'sanitation emergency'. TSC advocates a shift from high subsidy to low subsidy regime, greater household involvement, demand responsiveness, affordable toilets and promoting awareness. There is an urgent need to reach out to the poor and marginalized groups. Strategies adopted under the TSC include greater involvement of Panchayati Raj Institutions (PRIs) and Self-Help Groups (SHGs), developing schools and Anganwadis as vehicles for expanding the outreach of the programme, empowering women with knowledge and management skills, and undertaking research and development for technology, design and maintenance. Improvement of environmental sanitation is far more difficult task than building sanitary infrastructure. In view of the poor state of sanitary facilities and inadequate hygiene awareness, physical infrastructure and awareness building are equally important. The experiences of some of the successful sanitation projects, as illustrated earlier in this paper, clearly show the need for community participation and a sense of ownership by the users. Until the need for and benefits of sanitation are realized and internalized, mere provision of facilities will not provide the expected results.

Note

1. For a review of public health and sanitation in India under British rule, refer Ramasubban (1982).

References

Bartone, Carl, Janis Bernstein and Fredrick Wright (1990): "Investments in Solid Waste Management. Working Paper. No. 405". Washington DC, The World Bank.

Cairncross, Sandy (1992): "Sanitation and Water Supply: Practical Lessons from the Decade", Water and Sanitation Discussion Paper Series, No. 9. Washington DC: The World Bank.

– (2003): "Sanitation in the Developing World: Current Status and Future Solutions", *International Journal of Environmental Health Research*, 13 (June): S123-S31.

Census of India (2001): "Tables on Houses and Household Amenities and Assets. Series 1", Government of India: Office of the Registrar General and Census Commissioner.

– (1991): "Housing and Amenities. Occasional Paper No. 5 of 1994", Government of India: Office of the Registrar General and Census Commissioner.

Cooper, Kenneth J. (1997): "Human Waste Overwhelms India's War on Disease", *Washington Post Foreign Service*, February 17: A27.

CPCB (2000): "Status of Municipal Solid Waste: Generation Collection Treatment and Disposal in Class 1 Cities", New Delhi, Central Pollution Control Board.

DDWS (2002). "Sanitation", Government of India. Website: www.ddws.nic.in

DFID (2002): "Igniting Change: Tackling the Sanitation Challenge", Website: www.dfid.gov.uk

Government of India (2004): "Sanitation – Success Story", Website: ddws.nic.in

– (2003): "Towards Total Sanitation and Hygiene: A Challenge for India", Website: www.ddwsnic.in

– (2000): "Manual on Municipal Solid Waste Management", Ministry of Urban Development, New Delhi

Hasan, Arif (1988): "Low-cost Sanitation for a Squatter Community", *World Health Forum,* 9 (3):361-4.

IRC (International Water and Sanitation Centre). (2003): Website:www.irc.org location to sanitation.

ISEC (2005): "Restructuring Local Environment Management for Better Health Outcomes: Towards a District Level Pilot in Karnataka", Project Report, Bangalore, Institute for Social and Economic Change.

IIPS and ORC Macro (2001): *National Family Health Survey-2 (NFHS-2)-India, 1998-99*, Mumbai, International Institute for Population Sciences.

Krishnakumar, Asha (2003): "A Sanitation Emergency", *Frontline*, 20(24): 1-5.

Kundu, Amitabh (1991): "Microenvironment in Urban Planning: Access to Poor to Water Supply and Sanitation", *Economic and Political Weekly*, xxvi, 37.

Kurup, K. Balachandra (1996): *The Community-Managed Sanitation Programme in Kerala: Learning from Experience,* The Hague: IRC International Water and Sanitation Centre.

Listorti, James and Fadi M. Doumani (2001): "Environmental Health: Bridging the Gapes", World Bank Discussion Paper, No. 422. Washington DC: The World Bank.

Moraes, L.R.S., J.A. Cancio, S. Cairncross and S. Huttly (2003): "Impact of Drainage and Sewerage on Diarrhoea in Poor Urban Areas in Salvador, Brazil", *Transactions of the Royal Society of Tropical Medicine and Hygiene*, 97 (2): 153-8.

Nanan, D.F., White, I. Azam, H. Afsar and S. Hozhabri (2003): "Evaluation of a Water, Sanitation, and Hygiene Education Intervention on Diarrhoea in Northern Pakistan", *Bulletin of the World Health Organisation*, 81 (3): 160-4.

Narayan, Deepa (1995): "The Contribution of People's Participation: Evidence from 121 Rural Water Supply Projects". *Environmentally Sustainable Development Occasional Series 1,* Washington DC: The World Bank.

NSSO (1998): 52nd Round, *Morbidity and Treatment of Ailment,* New Delhi: National Sample Survey Organisation, Government of India.

– (1999): 54th Round, *Drinking Water, Sanitation and Hygiene in India*, National Sample Survey Organisation: Government of India.

Pasha, Hafiz and Michael G. McGarry (eds.) (1989): "Rural Water Supply and Sanitation in Pakistan: Lessons from Experience", World Bank Technical Paper No. 105, Washington D C: The World Bank.

Pathak, Bindeshwar (1997): "Cost Effective and Appropriate Sanitation System", *Nagarlok*, XXIX (4): 78-90.

Ramasubban, Radhika (1982): "Public Health and Medical Research in India", *Sarec Report R 4: 1982.* Stockholm: Swedish Agency for Research Cooperation with Developing Countries.

Richards, Edward P. (2002): "The Role of Medical and Public Health Services in Sustainable Development", Website: http://www.eli.org

Roy, M.N. (1996): "An Alternative Approach for Sustainable Development of Rural Sanitation – The West Bengal Experience", *Journal of Rural Development*, 15 (1): 119-39.

Sara, J. and Travis Katz (1998): *Making Rural Water Supply Sustainable: Report on the Impact of Project Rules,* Washington DC: The World Bank.

Sekher, T.V. (2003): "Sensitising Grass Roots Leadership on Health Issues: Experiences of a Pilot TV Project", *Economic and Political Weekly*, XXXVIII (46): 4873-79.

Serageldin, Ismail (1994): *Water Supply, Sanitation, and Environmental Sustainability*, Washington DC: The World Bank.

Shi, Anqing (2000): "How Access to Urban Potable Water and Sewerage Connections Affects Child Mortality", Policy Research Working Paper, No. 2274. Washington DC: The World Bank.

UN-HABITAT (2003): *Water and Sanitation in the World's Cities: Local Action for Global Goals,* London: Earthscan Publications Ltd.

United Nations Centre for Human Settlements (UNCHS) (1997): "Monitoring Progress on Implementation of the Habitat Agenda One Year after Habitat II", *Habitate Debate*, 3 (3).

WHO (2002): *The World Health Report 2002*, Geneva: World Health Organisation.

– (2001a): Water-related Diseases. Website: www.who.ch Related to Water and Sanitation

– (2001b): www.who.org, related to Action Against Infection, A Newspaper and Hygiene Link for WHO and its partners, 2 (1).

– (1995): WHO Expert Committee on Environmental Sanitation, World Health Organisation Technical Report Series 10,5, WHO, Geneva.

WHO/UNICEF (2000): *Global Water Supply and Sanitation Assessment 2000 Report.* 2000. Geneva/New York: WHO/UNICEF.

Wisner, B. and J. Adams (eds.) (2002): Environmental Health: in Emergencies and Disasters. Geneva: World Health Organisation.

WSP-SA (1999): Website: www.wsp-sa.org related to Decentralised RWSS Management Gram Panchayats vs Village Water and Sanitation committees.

– (2001): Website: www.wsp-sa.org, related to Why Some Village Water and Sanitation Committees are Better than Others: A Study of Karnataka and Uttar Pradesh (India).

– (2002): Website: www.wsp-sa.org related to Making Sanitation Work

– (2002): Wesbsite: www.wsp-sa.org, Implementing Sector Reforms: A Review of Selected State Experiences.

– (2003): Website: www. wsp-sa.org Related to Sanitation

WSSCC (2000): Website: www. wsscc.org Related to Water and Sanitation.

World Bank (1992): *World Development Report 1992. Development and the Environment*, New York: Oxford University Press.

– (1993): *World Development Report 1993. Investing in Health*, New York: Oxford University Press.

– (1999 a): *Rural Water Supply and Sanitation.* New Delhi: The World Bank.

– (1999 b): *Urban Water Supply and Sanitation.* New Delhi: The World Bank.

World Health Organisation (2004):. Water, Sanitation and Health. Electronic Library: A Compendium of WHO Information on Water, Sanitation and Health (as cited by Chatterjee).

III Urban Environment and Disaster

10 Urban Ecological Issues with Special Reference to Mumbai

K. Sita and Malabika Ray

The massive deluge of July 26, 2005 in Mumbai brought urban ecological issues into sharp focus. Generally, they tend to be overshadowed by economic, social and infrastructure factors. This is evident when one peruses the developmental plans drawn up from time to time for Mumbai.

The ecological issues that come to the fore in urban areas can be broadly classified into three groups, i.e., universal, regional and specific. Universal issues arise out of the intrinsic character of urban areas. Regional are those that are prevalent due to the characteristics of the area where the city is located. The specific are due to the distinctive features of the urban centre.

Universal issues, found in all urban centres are primarily the result of the concentration of population and non-agricultural activities in these nodes. The land use pattern is modified and built-up areas and paved surfaces are predominant. The seepage of water into the ground is reduced resulting in higher run-off. The ecological services needed to serve the concentration of population are of large magnitude and cannot be met from local resources. It has given rise to ecological appropriation by large cities, which is referred to as the ecological footprint of a city. It is well illustrated by the distant sources tapped for water supply to meet the demands of a city and areas utilized as sinks of waste generated in the city.

The regional issues are a result of the characteristics of the macro and meso region in which the city is located. For cities in developing countries, one finds that due to the higher incidence of poverty, a lower priority is given to ecological issues. In tropical countries, the higher temperatures result in a more rapid putrification of wastes, which necessitates their collection and disposal more frequently. In tropical monsoon climates, the rainfall is concentrated in a few months and its intensity is high on some days. The storm water drainage network may be unable to cope up with the high run-off leading to flooding.

The specific issues are due to local relief, the site characteristics, the nature of a city's economy, etc. For example, Mumbai was initially a cluster of islands off the west coast of India. It acquired its present form due to reclamation over a long period of time. The suburban area has a central group of hills and is occupied to a large extent by the Sanjay Gandhi National Park. Streams rising in these hills and flowing to the sea were the drainage channels. One such important waterway is the Mithi river flowing into the Mahim creek. Extensive areas along the coast are low-lying and occupied by mangroves, marshes, mud flats, saltpans, etc. Unbridled urbanization has led to large-scale encroachment into these ecologically fragile areas. The consequence was the recent flooding on an unprecedented scale that resulted in loss of life, immense human suffering and economic loss.

The danger inherent in the neglect of ecological issues in urban areas was brought into sharp focus on July 26, 2005 when the suburbs of Mumbai experienced an unprecedented rainfall. The resultant deluge caused a heavy loss of life and property in parts of the metropolis and disrupted normal life for several days. Though authorities tried to explain it as due to the spell of unusually heavy rainfall concentrated in a short span of time, it was evident that it was partly due to neglect of environmental factors in the planning of the city's growth. This is reflected in the various development plans drawn up for Mumbai from time to time over the last 60 years from Meyer and Modak in 1945 to MMRDA Draft Regional Plan for 1996-2011. In all the plans the economic, social and infrastructure imperatives tended to overshadow the environmental factors. It must be borne in mind that the latter are also important as an urban area can survive only if there are reliable supplies of water, energy, food, raw materials, etc. Equally important are the sinks for the wastes generated in a city. It was to focus attention on these aspects that Rees and Wackermagel (1994) put forward the concept of the 'ecological footprint' which can be applied at different scales of analysis, whether to a city, a country, a region or the earth. The concept draws attention to the significance of the interrelationship between man and environment which has to be given due importance if urban centres with their increasing populations are to evolve in a sustainable manner.

Urban Ecological Issues: A Typology

The urban ecological issues that come to the fore in any urban centre can, for purposes of convenience, be broadly classified into three groups: universal, regional and specific. Universal issues arise out of the intrinsic character of urban centres, i.e., that they are areas of concentration of people and of economic activities. This gives rise to a drastic modification of the original environment. The nature of the attendant problems might vary from one city to another; however, the issues are of relevance in cities of both the developed and the developing countries. Since developed countries have already experienced the problems and tried to find solutions, these could be adapted to suit the requirements of cities in the developing countries.

The regional issues in any city are due to the characteristics of the macro- and meso-region in which the city is located. In this case only the experience of other cities of the region can be adapted and hence they pose a greater challenge to policy makers and administrators in Third World cities.

The specific issues are due to the distinctive ecological characteristics of an urban centre. They arise out of its site characteristics, location, etc. The solutions in this case have to be city specific, based on an analysis of its unique features.

Universal Issues

As pointed out in the earlier section, they arise out of the intrinsic characteristics of urban centres.

1. Due to the concentration of population in a small area, the surface features are modified. Buildings and roads replace open spaces. This gives rise to:
   (a) Reduction in the seepage of water into the ground, and hence an increase in surface flow. The natural drainage system, consisting of rivers, rivulets, holding ponds, etc., gets disrupted due to construction (Cover Story, 2005). The amount of water to be handled by the storm water drainage network (SWDN) during heavy rains increases.
   (b) The concrete and asphalt surfaces absorb more heat, giving rise to the urban heat island effect. In extreme cases, it may result in modification of the pattern of airflows.

2. The concentration of population and of economic activities gives rise to a heavy demand for various resources. It is pertinent here to consider the ecological resources. A good example is water supply. Since resources of water on such a large scale are rarely available within the urban area and at the same time it is not feasible to transport water over long distances, the city generally depends on its immediate hinterland for supplies to quench its thirst. An exception is Singapore, which obtains part of its water from Malaysia. In Mumbai, initially the three lakes of Vihar, Tulsi and Powai on the Mithi River in Salsette were tapped. By the end of the 19th century, they were inadequate to meet the growing demand and the Tansa scheme was undertaken. By the mid-20th century, a further augmentation was through the Vaitarna scheme and later the Upper Vaitarna and the Bhatsai. At present, since supply still falls short of demand, the middle Vaitarna project is being undertaken. Thus, to meet the increasing demand of the metropolis, water is brought from distances of over 150 kilometres (Phadke and Rao, 1994). This naturally deprives the areas in south Thane district of their water and illustrates the ecological appropriation by cities. Again, for the supply of building material such as bricks, road metal, gravel, etc., the dependence of the city on its immediate hinterland has given rise to quarrying which not only defaces the landscape but also affects flow rates of water, susceptibility to landslides, etc.
3. While the city depends on its hinterland for the supply of certain ecological services, it also uses it as a sink for some of the wastes produced in it. A mega city such as Mumbai generates a vast quantity of solid, liquid and gaseous wastes. The dumping grounds of solid wastes are generally initially located away from the residential areas. However, as the city expands, it tends to engulf them and developing new areas is a problem as there is opposition from residents who do not want them in their backyard Similarly, the collection, treatment and disposal of liquid wastes are major tasks for civic authorities. It is estimated that Mumbai generates 2,800 million litres of sewage daily; probably it is higher. The present sewerage network of 1,400 kilometres covers only 60 per cent of the city; in the remaining areas it is discharged into nullahs, rivulets etc. The city is divided into 7 sewerage zones each having a major pumping station. The

sewerage is partly treated using aerated lagoons at Versova, Ghatkopar and Bhandup. It is later disposed off 3 kilometres away from the coast through pipelines running along the seabed; these are the marine outfalls.

The large quantity of gaseous wastes is due to the concentration of industries, vehicular traffic, etc. These give rise to problems of air pollution, which may at times become aggravated and a cause for concern. For example, during winter, when temperature inversions are common in the mornings, the dispersal of pollutants is affected giving rise to the incidence of smog. It results in high incidence of respiratory ailments. It also reduces visibility and may cause disruption of transport. In Delhi it has necessitated a strict monitoring of the fuel used by vehicles and a shift to CNG, a cleaner fuel.

The above are some of the issues in all cities; some have been more successful in tackling them than others. For example, Singapore has a three-stage treatment for liquid wastes that produces an effluent so clean that it is being used by the semi-conductor industry on an experimental basis. The controls on air pollution have resulted in its being below international standards.

In contrast, Kolkata has no large sewage treatment plant (Stout, 2003). Part of the sewage is channelized to the wetlands. In the past fishermen used small gates to direct it into small ponds. The region's ample sunlight and aeration converted the organic matter into food for plankton while the vegetation in the wetlands could clean the water of heavy metals and bacteria. The cleaned water was then released to neighbouring maize and vegetable fields. The system was effective and the sewage supported a large agricultural economy. However, the wetlands have shrunk due to urbanization and the system is unable to cope with the increase in domestic and industrial wastes.

Regional Issues

These issues, as pointed out earlier, are due to the characteristics of the region in which the city is located.

1. Cities in developing countries differ from those of the developed world in that the incidence of poverty is greater in them. This leads to a large proportion of the population living in informal housing or slums. For example, in Mumbai nearly 50 per cent of the population lives in the slums. It is but natural, that as a consequence, economic issues are given a priority. This is a trait

common to cities in this region such as Dhaka, Bangkok, etc. Hence, in urban planning in these cities, ecological factors are overshadowed by other more pressing concerns.

2. Cities in tropical countries have higher temperatures as compared to those in temperate countries. This results in rapid decomposition of wastes and necessitates more frequent collection and disposal. Since many of these countries are also part of the developing world, solid waste collection and disposal places a heavy strain on the urban economies. In many cases they are inadequate.
3. In cities having a tropical monsoon climate such as Mumbai, the rainfall is concentrated in a few months. The intensity of rainfall is very high on some days. The quantum of water to be drained by the SWDN may exceed the capacity of the network. In Mumbai, the problem is accentuated if the spell of heavy rainfall coincides with a period of high tide. It results in flooding of low-lying areas.

Problems of drainage are common to many cities located near the coast in low-lying areas. For example, in Bangkok the network of canals is part of the natural drainage system; however, the filling up of several canals for providing roads has created drainage problems. The situation has been exacerbated due to squatter communities building their houses along the canals; they are known as 'klong' settlements. It has impeded the maintenance of the canals. In Kolkata, the Hooghly River provides the main drainage system. However, the slope of the land to the east has led to the creation of the eastern Kolkata wetlands. The Salt Lake, one of the largest of the wetlands, acted as a cushion against floods. The increase in built-up area has eaten into the wetlands and increased susceptibility to flooding.

Thus, the regional issues which cities in the tropical monsoon belt have in common, if they are coastal cities, are problems of drainage during spells of heavy rainfall and disposal of wastes, which have to be collected daily.

Specific Issues

The specific issues are due to the distinctive characteristics of a city. For example, Mumbai is a coastal city and, as is well known, it originally comprised a group of islands off the west coast of India. It acquired its present form due to reclamation over the last 300 years. Hence, in the city almost 40 per cent of the area comprises reclaimed land.

Moreover, due to two parallel ridges of low, broken hills on the east and west, drainage is towards the low-lying intervening area between them, making it prone to flooding during the monsoon. In the suburbs in Salsette to the north, the coastal areas are fringed by extensive stretches of tidal marshes, swamps, mangroves, etc. Salsette has a hill core in the north towards the centre and the drainage pattern is a radial one from the hills towards the coast (Figure 1). The main streams are the Mithi and Dahisar draining into the Mahim and Manori creeks respectively. Since the suburbs have accommodated most of the population growth of Mumbai in the last 50 years, it has led to indiscriminate modification of the natural landscape (Srivastava and Mukerji, 2005). This can be illustrated with reference to the Mithi river. It originates in the central hills of Salsette. The three lakes of Tulsi, Vihar and Powai, the earliest sources of water for Mumbai, are on it. From Powai Lake, it flows 15 kilometres before joining the Arabian Sea at Mahim creek. There are a lot of mangroves near its mouth and a bird sanctuary. The increase of built-up land in the suburbs due to the rapid growth of population has affected the drainage channel (Arunachalam, 2005). Many factors have led to the reduction in the flow of the Mithi river:

- There has been a narrowing of the channel at places due to encroachments. For example, the construction of the airport altered the course and the natural gradient while the second runway built in the 1970s literally squeezed the river (Sekhar, 2005c).
- The development of the Bandra Kurla Complex (BKC) involved the reclamation of extensive wetlands near the estuary (Samant, 2002).
- The proliferation of slums and small industrial units along its banks has led to heavy discharge of effluents and sewage. In fact, the Environment Status Report of Mumbai, 2002-3, refers to it as "one of the heavily polluted storm water drains".

Naturally, there is a tendency of flooding during spells of heavy rainfall, particularly if they coincide with the period of high tide.

Hence, in low-lying coastal cities, such as Mumbai, located in the tropical monsoon belt, an important ecological issue is that of maintenance of wetlands and of natural drainage channels (ToI, 2005).

Figure 1
Salsette: Relief and Drainage

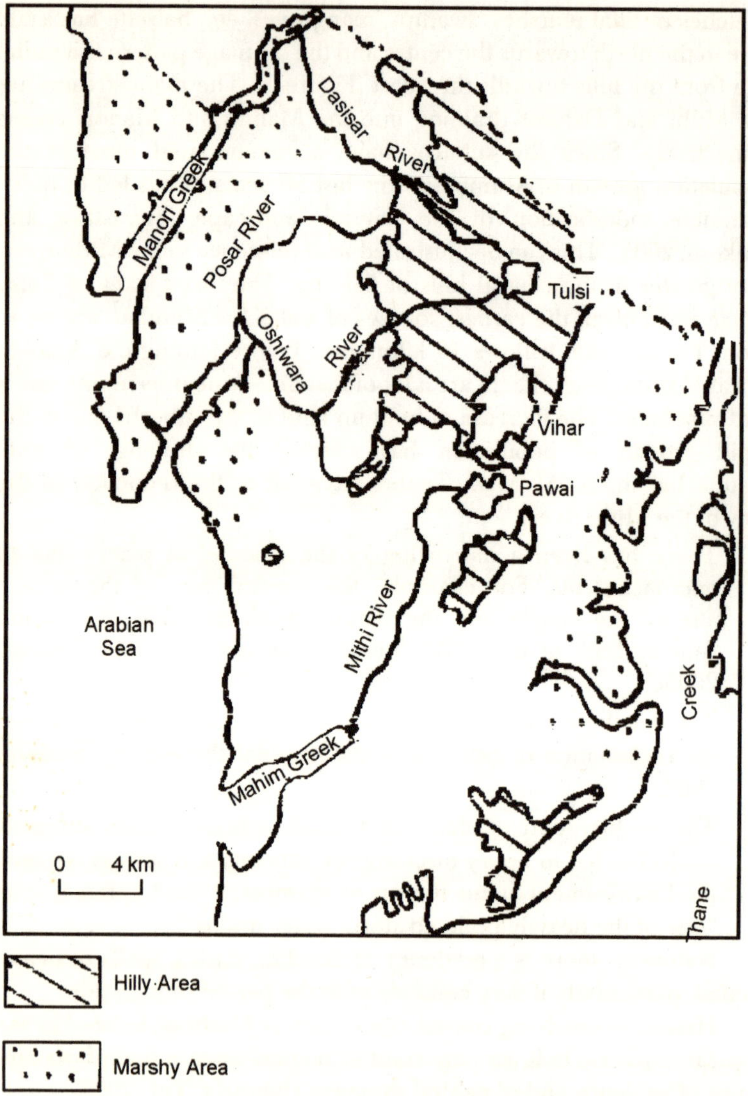

Conclusions

The growth of population and increasing urbanization are leading to large concentrations of population on limited areas; this is inevitable. However, the problems could be reduced if better cognizance is taken of the physical environment, so that environmentally vulnerable areas such as the wetlands are protected (Sekhar, 2005 a and b). In the case of Mumbai, it is evident that a universal issue, i.e., increase in surface flow of water during spells of rainfall has become accentuated due to the regional characteristic of concentration of rainfall in a few months in the tropical monsoon belt in which it is located. The specific feature of its coastal site in which reclamation played a major role in enhancing the land resources has hence led to drainage channels being of critical importance. The neglect of these has led to problems of flooding.

It is pertinent to conclude with a story from our scriptures. When Vamana went to King Bali's court he was asked what boon he wanted. Vamana requested for an area equivalent to three of his footprints. When King Bali agreed to his request, Vamana grew in size. He occupied the entire land surface with one footprint, and the whole ocean surface with the second. Since there was no space for the third, the king asked Vamana to place his foot on his head. With this move, the king was pushed into oblivion (Pathala loka). This illustrates the danger that cities face if cognizance is not taken of environmental factors and efforts made to reduce the size and magnitude of ecological footprints.

References

Arunachalam, B. (2005): "Drainage Problems of Brihan Mumbai", *Economic and Political Weekly*, September 3, pp. 3909-3911.

Cover Story (2005): "Mumbai Maimed", *Down to Earth,* 14(9): 24-34.

Phadke, V.S. and Rao, K.V.S. (1884): "Domestic Water Supply to Greater Bombay", in N.L. Gupta (ed.), *Urban Water Supply,* Rawat Publications, Jaipur and Delhi.

Rees, W.E. and Wackermagel, M. (1994): "Ecological Footprints and Appropriated Carrying Capacity", in Jansson, A.M. et al. (eds.), *Investing in Natural Capital*, Island Press, Washington D.C., pp. 362-390.

Samant, H.P. (2002): "Quantifying Mangrove Cover Change in and around Mumbai using Satellite Data", *Proc. of the National Seminar on Creeks, Estuaries and Mangroves*, pp. 334-337.

Sekhar, V.C. (2005a): "See How the City Changed", *Times of India*, August 3, p. 1.

– (2005b): "From Wetlands to Bad Lands", *Times of India*, August 4, p. 2.

– (2005c): "How the Airport ate up Mithi River", *Times of India*, August 7, pp. 1-2.

Srivastava, S. and Mukerji, D. (2005): "Emerging Cityscapes and Environmental Issues", *Economic and Political Weekly*, September 3, pp. 3905-3909.

Stout, K.L. (2003): "Global Challenges", *CNN*, June 15, (http//: transcripts. Cnn.com/TRANSCRIPTS/6306/15/gc.00html.

Times News Network (2005): "Nature's Revenge", *Times of India*, August 2, p. 1.

11 Cyclones in Andhra Pradesh: Damages and Response

U.V. Somayajulu

India, the world's second most populous country, is affected year after year by inevitable disasters – floods, cyclones, droughts, earthquakes, etc. About 40 per cent of the country's population live in poverty and these poor sections of the society are the most affected group in case of any disaster. About 6 per cent of the population is affected by disasters and in 1998, 37 per cent of the area was identified as flood prone National Centre for Disasters Management (NCDM). The 7,516 km long Indian coastline is the earth's most cyclone battered stretch.

It is estimated that about 44 per cent of Andhra Pradesh's territory is vulnerable to tropical storms and related hazards such as high winds, storm surges and floods.

This paper presents a brief history of cyclones in Andhra Pradesh and discusses the damage caused by these cyclones in terms of loss or property and human life and displacement, on the basis of secondary data and government documents. The paper also presents the findings of a study carried out among the households living in the coastal villages of the state in terms of cyclone early warning information, cyclone contingency plan, mitigation facilities, etc.

The study indicates the need to improve the cyclone early warning mechanism and prepare cyclone contingency plan and provide mitigation facilities in the vulnerable villages.

Cyclones

Cyclones are such ubiquitous disaster phenomenon that they go by many names the world over, so understanding their nature would help minimize the destruction. Cyclones are global phenomenon and go by many names, viz., tropical cyclones in India, cyclones and hurricanes in the West and Willie-willies in Australia (Reddy et al., 2000).

Tropical cyclones which are the bane of the warm oceans, affect thousands of square kilometres of ocean surface and the lower atmosphere. A cyclone's perimeter measures 1,000 km but its actual powerhouse is located within its 100 km radius.

There are two meteorological imperatives for a cyclone to occur – the sea surface temperature has to be more than 26.5^0 C and the formation of cyclone requires an anticyclone, a body of moving air with higher pressure than the surrounding air in the upper troposphere.

Movement of Cyclones

Tropical cyclones follow a basic parabolic pattern but they are also known to behave erratically and change direction without warning. The cyclones travel over oceans at a speed of 16-24 kmph, but once they hit land, they move inland with destructive force of 160 km or more depending on the original size of the storm.

Storm Surges

The storm surges (tidal waves) are caused by a combination of the raising of the sea level by wind driven water and low pressure at the eye of the cyclone. Surge height (up to 7 metres) and length (up to 50 km) depends on the tides, the rate of water run-off from the land, onshore winds, and coastal contours and configuration. Surges cause the most havoc when they recede, sucking away land, property, trees and living beings with the force of their retreat. On an average, storm surges kill seven times more people and damage three times more crops than severe but surge free cyclones.

Wind

Cyclonic winds – in linear moving as well as rotary forms – cause massive damage to houses, agriculture fields and infrastructure.

Rain

Cyclones are usually accompanied by heavy downpours, measuring anything from 250 cm.

Role of Global Warming

Global warming, which is caused by the increasing quantum of green house gases such as carbon dioxide, raises sea surface temperature and exacerbate cyclonic frequency, intensity, duration and track modification. Global warming also raises sea level through continuous and irreversible melting of polar ice caps, eventually leading to inundation of chunks of coastal lands.

Classification of Storms

Table 1 gives the international classification of storms ranging from depression to severe cyclonic storms.

Table 1
Classification of Storms

Type of Storm	*Wind Speeds*	
	Km/hr	*M/sec*
Depression	30-50	4.7-7.5
Deep depression	50-65	7.8-8.2
Cyclonic storm	65-90	9.4-13.0
Severe cyclonic storm	90-120	13.3-17.5
Severe cyclonic storm (with core of hurricane winds)	Above 120	Above 17.8

In the subsequent sections, the situation in India and Andhra Pradesh (AP) is discussed in terms of cyclones, damage caused and the response of the government to these cyclones. The findings of the household survey (carried out in 10 coastal districts of AP) are also discussed in terms of cyclone early warning systems, contingency plans and mitigation facilities.

The Situation in India

Disasters in India

India, the world's second most populous country, is the world's most geographically, topographically and demographically diverse country, comprising some of its hottest and coldest spots, its wettest valleys and driest deserts, dealing with myriad disasters.

Table 2 gives the details about disasters in India in terms of area and population effected.

Table 2

Disasters in India

	%
Population affected by disasters	6
Earthquake-prone area	57
Flood-prone area	37
Drought-prone area	28

Frequency of Cyclones

The frequency of cyclones varies in various parts of the world, but 7,517 km long Indian coastline is the earth's most cyclone battered stretch, and staggers incalculable annual losses of lives and property. Though tropical cyclones in the Bay of Bengal and the Arabian Sea are the least frequent and usually of moderate intensity, the Indian sub-continent remains the worst cyclone affected region in the world.

In the century between 1877 and 1977, 377 cyclones crossed the east coast of India and Bay of Bengal between the latitude of 8 degree N and 22 degrees N. About a third of these cyclone caused landfalls on the northern Orissa and Bengal coasts between 20 and 21 degrees N. The Krishna delta in Andhra Pradesh at the latitude of 15-16 degree N is the next most cyclone prone region. The most vulnerable districts in Andhra Pradesh are Nellore, Krishna, Srikakulam and East Godavari.

Table 3

Major Cyclones that Hit India

State	*Period*	*Population affected*	*Loss of lives*	*Crop affected*	*Houses*	*Public utilities*
Andhra Pradesh	Nov 14-20, 1977	7.1M	8000+	3.6MH	1.01M	Rs 1776.6 M
Orissa	June 4, 1982	7.32 M	243	1.58 MH	8.19L	Rs 5591 M
Andhra Pradesh	May 5-9 1990	7.78 M	928	45000HC	1.4M	-
Andhra Pradesh	Nov 6, 1996	-	1077	5.11LHC	6.09L	Rs 5591M
Gujarat	June 9 1998	-	1308	10,000 HC	10,000 + 5000 (partly)	-

M = Million; L = Lakhs; - = Not Available HC= Hectares; MHC = Million Hectares

Major Cyclones in India

Of the Indian's coastline, east coast is the most cyclone vulnerable area due to geographical and demographic reasons. Thickly populated coast of Andhra Pradesh bears the brunt of cyclones. Table 3 gives the details about the major cyclones that hit India.

The Situation in Andhra Pradesh

Population and Area

Andhra Pradesh is the fifth largest state in India, in terms of area as well as population. Spread over 2.75 lakh sq km, it comprises 8.4 per cent of the country's total geographic area. Andhra Pradesh can be divided into three geopolitical divisions – Telangana, Rayalaseema and coastal Andhra. Table 4 gives the area and population of these three regions. The coastal Andhra region accounts for one third of the land area and two-fifths of the state's population.

Table 4
Areas and Population of Andhra Pradesh by Region

Region	*Area (sq. km)*	*Population (2001)*
Coastal Andhra	92906 (34 %)	31570722 (42%)
Telangana	114840 (42%)	30696566 (41%)
Rayalaseema	67299 (24%)	13460253 (17%)
Total	275045	75727541

Coastline

Andhra Pradesh has a long coastline stretching approximately 1,030 kms and equally long history of cyclones of varying intensity over the past 100 years. The total number of villages perched along the coastline is 2,482. Of these, 500 are within 5 kms swathe extending inside from the coast, 601 are within 5-10km and 1,381 villages are between 10-20 km.

Coastal Andhra Pradesh covers an area of 92,906 sq kms with a total population of 31,570,722. The details of coastline by districts are furnished in Table 5.

The land on either side of the Godavari and the Krishna rivers consists of extensive, flat delta plains leveled between 4-6 feet above mean sea level. These areas are most vulnerable to storm surges.

Table 5

Details of Coastline by District

District	*Length of coastline (km)*
Srikakulam	182
East Godavari	177
Nellore	174
Visakhapatnam	155
Prakasam	116
Krishna	122
Guntur	62
Vizianagaram	25
West Godavari	17
Total	1030

The coastal districts are primarily agricultural – Guntur, Krishna, East Godavari and West Godavari have highly productive network of irrigation canals founded on the Krishna and Godavari rivers. These four districts, along with Srikakulam, also face the brunt of floods during the monsoons. The coastline ranging from Srikakulam in the north and Nellore in the south is affected by at least one cyclone every year. The entire state suffers the consequences of cyclones because each district except Vizianagaram, West Godavari and Guntur, has one or more faces with over 100 km of very vulnerable seacoast.

The area immediately close to the coast is relatively less populated, less accessible due to poor communication. The people living in the belt of 20 kms from the coast generally comprise fisher folk and weaker sections and majority of them live in thatched houses which are vulnerable to the wind pressures of cyclonic gale which exceed 100 kms per hour.

History of Storms in the State

The state of Andhra Pradesh experienced 71 cyclones during 1892-1997, each time with colossal damage. Between 1892 and 1977, 56 cyclones affected the seven coastal districts. The district-wise figures given in Table 6 indicate that Nellore, Krishna and Srikakaulam districts had more number of cyclones.

Table 6

No. of Cyclones by Districts (1892-1977)

District	*No. of Cyclones*
Nellore	15
Krishna	14
Srikakulam	10
East Godavari	8
Prakasam	6
Visakhapatnam	2
West Godavari	1

Factors Responsible for Vulnerability

Some of the factors responsible for vulnerability of the state to cyclone are:

- Almost half of the storms in the bay of Bengal become severe cyclones often accompanied by storm surges
- Low-lying areas along the coast are vulnerable to extensive flooding and deep inland sea water incursion
- High concentration of population, infrastructure and economic activities along the coast
- Lack of proper maintenance of the flood protection and irrigation systems, drains, embankments etc.
- Lack of comprehensive coastal zone and delta management

The Three Deadly Cyclones

In the past two decades, three major cyclones caused immense loss of human lives and livestock and massive damage to property, both of people and the government, viz.,

- November 1977
- May 1979
- November 1984

The November 1977 cyclone was the deadliest with 9,941 people dead, 900 people missing and 34 lakhs rendered homeless. Cattle and livestock losses stood at Rs 2.5 lakhs, crops were lost in 33,36,000 acres and 1,014,800 houses were damaged. Losses to buildings and public property were estimated at Rs 172 crores.

Table 7

Details of the Estimated Losses and Damages

Year	*Districts effected*	*Human deaths*	*Loss of livestock*	*Houses damaged*	*Crop damaged (lakh ha.)*	*Crop Loss (Rs in lakh)*	*Revenue exp (Rs.in lakh)*
Nov 1977	8	992	431786	1014800	33.36	1720.00	5305.80
May 1979	10	638	25082	609400	0.73	18000.00	7814.72
Nov 1984	4	575	90650	320000	2.07	9489.93	4928.89
Nov 1985	7	16	0	3196	1.06	2426.12	1850.54
Nov 1987	12	119	0	110553	9.61	12648.77	4141.65
Nov 1989	5	69	7117	149112	0.62	4082.37	10015.59
May 1990	14	976	5170301	1439659	4.80	224716.00	13357.78
Nov 1994	7	172	512	79220	3.97	62593.18	2968.70
Nov 1996	3	1077	19856	609628	5.11	226857.00	30822.00
Total		13563	5745304	4335568	87.15	578073.37	81205.67

Impact of Cyclones

The major impact of cyclones can be broadly categorized as below:

- Loss of lives, injuries and other health consequences such as epidemics, and post traumatic stress disorders (PTSD)
- Loss of habitat
- Loss of cattle and damage to crops and agricultural fields
- Damage to public utilities
- Disturbance and damage to the ecosystem

Most lives are lost during a cyclone on account of floods and the devastating storm surge that often accompany cyclones. In case of severe cyclonic storms with storm surges, more than 90 per cent of the fatalities occur due to drowning, either during the incoming water phase or during the out surges. In severe cyclonic storms without storm surges, the deaths are more or less evenly divided between drowning and the collapse of buildings.

The most common health problems include water borne diseases such as diarrhoea, dysentery, typhoid, viral hepatitis, respiratory diseases such as pneumonia and whooping cough, measles, gastroenteritis, cholera, conjunctivitis, and fever.

Response Mechanism

The various phases of disaster response mechanism comprises of the following:

- Early warning
- Evacuation
- Rescue
- Relief
- Rehabilitation
- Development

The three 'R's, viz., rescue, relief and rehabilitation are important in the context of any disaster. Explanation of these is as follows:

1. *Rescue:* Save or set free from danger or harm
2. *Relief:* Alleviation or deliverance from pain, distress and anxiety
3. *Rehabilitation:* Restore to effectiveness or normal life by training, etc.

Early Warning

In view of availability of satellite images and remote sensing, cyclone forecasting has become more sophisticated and accurate, but the dissemination of warning percolates down to the lower levels rather late, mainly due to inadequate communication network and difficulty in reaching the inaccessible villages.

Evacuation

Forecasting plays a crucial role in facilitating evacuation. Though it is regarded as a costly and difficult method, more often it becomes inevitable. Refusal of people to leave houses, fields and lack of initiative from officials hampers the evacuation process.

Rescue and Relief

This 'sandwich' phase usually lasts for 2-4 weeks, and involves the services of armed forces, government agencies, NGOs, etc. The political and bureaucratic factors influence this phase.

Rehabilitation and Development

Rehabilitation is a long-term slog and it should ideally include restoring life sustenance systems as rebuilding the social fabric of the affected communities. The rehabilitation programme should take care of health, education, special needs of women and children, farmers, agricultural labourers, artisans, marginalized and vulnerable groups. Several factors including the political considerations determine the ex gratia payments extended to the affected people. In 1996 cyclone, the maximum ex gratia payment given was Rs 1 lakh.

Gender Issues

By virtue of their lower economic, social and political status and position, women tend to be among the most vulnerable to disasters and to be held responsible for the family reconstruction after the disaster. For example, the allotment of house sites for the construction of houses has invariably been in the name of husband or son. Many cases of 1977 cyclone affected women (who lost men in the family or from single women healed households) not having a shelter of one's own, even after 2-3 decades provide evidence for this.

Health Care

Health care must take priority over everything else in the aftermath of a cyclone, as every aspect of disaster management from food to medication to accommodation is related to health.

Relief and Rehabilitation

1977 Cyclone

Ex gratia payment was Rs 1,000 for every deceased person (including children of 6+ yrs). The total amount paid across the Krishna, Guntur, Prakasam, West Godavari, East Godavari, Visakhapatnam, Srikakulam and Khammam districts was Rs 17.39 crores. The damages reported to the Union Government was estimated to be Rs 1,000 crores and the Central Government provided an assistance of Rs 56.52 crores. The union government also provided Rs 3.11 crores towards rehabilitation of tobacco farmers, small and marginal farmers, etc. Rice and wheat of 45,000 tonnes each was also supplied by Government of India (GOI). Rs 11.46 crores was sanctioned by GOI for irrigation, municipal development and Panchayat Raj.

The Relief and Resettlement (R&R) measures adopted include:

- Remission of revenue to the tune of Rs 7.34 crores
- Rs 1.61 crores towards fishing vessels, tackles and nylon nets for fishermen
- Rs 200 each paid to the 41,828 weavers
- Rs 68.5 lakhs distributed to the 7,500 artisans

1996 Cyclone

The details of relief extended to the Nov 1996 cyclone victims include:

- Ex gratia of Rs 100,000 to the families of deceased (Rs 5,000 for missing person, and Rs 95,000 on confirmation of death)
- Rs 10,000 for serious injuries and Rs 2000 for minor injuries
- Gratuitous relief of Rs 1000 + 200 and 25 kg of rice for fully damaged houses, Rs 500 and 25 kg of rice for partially damaged houses
- Rs 500 for damaged looms, Rs 200 for damaged yarn, and 25 kg of rice to weavers
- Rs 1,000 for lost boat, Rs 500 for damaged boat and Rs 200 for damaged net to fishermen
- Rs 625 per hectare for fully damaged crops, Rs 2,500 per hectare for removal of sand
- Rs 1000 for petty shops damaged in cyclone
- 3 litres kerosene to all affected families
- Postponement of water tax collection in the affected areas
- Postponement of recovery of crop loans in the affected area, sanctioning of fresh loans

Expenditure Incurred 1996-97 Cyclone

Table 8 gives the details of expenditure incurred by the Government of AP during the 1996-97 cyclones. The total expenditure during the 1996-97 cyclones was Rs 27,318 lakhs. Of this, about one-fifth was towards assistance for repairs to damaged houses, while about one-tenth each was spent towards ex gratia to bereaved families, Municipal workers, Panchayat Raj workers.

Table 8
Expenditure in 1996-97 Cyclone

Sector	*Expenditure (Rs lakhs)*	%
Medicines supply	117.77	0.43
Roads and Buildings Department	1467.47	5.37
Supply of seeds, fertilizers and agricultural inputs	2117.83	7.75
Irrigation and flood control works	2159.21	7.91
Panchayat Raj works	2667.61	9.77
Municipal works	3091.34	11.32
Electrical works	2182.04	7.99
APSIDC	40.00	0.15
Exgratia to bereaved families	3248.16	11.89
Food and clothing	2623.58	9.61
Assistance for repairs to damaged homes	5212.88	19.08
Supply of fodder	0.5118	0.00
Supply of seeds and fertilizers	118.02	0.43
Others	2267.71	8.30
Total	27318.00	100.00

Community's Perspective

The results of the household survey carried out in 50 coastal villages of Andhra Pradesh are discussed in this section to get an understanding of the community's perspective with regard to the various aspects relating to cyclones.

Profile

Table 9 gives the profile of the households in terms of socio economic and demographic characteristics. Illiteracy is an issue, as one-tenth of the households do not have even one literate member. This is evident from the comparison of average number of literates in the household with that of household size. This will have bearing on the cyclone early warning message dissemination.

Majority of the households belong to SC/ST and OBC groups and are economically insecure as they are mostly unskilled labour or fishermen. Two-fifths of the households own livestock and this implies that they need to ensure safety of these in the event of any disaster. In spite of experiencing maximum number of cyclones, Andhra Pradesh is well behind the national average in terms of safe shelter during cyclones. Most people live in kutcha houses, which are vulnerable to collapse during floods and cyclones.

Table 9

Profile of Households in Coastal AP

Education	%
Average No. of literates	2.8
Average No. of household size	5.0
% of Households with no literate	11
Caste	
SC/ST	23
OBC	63
Others	14
Occupation	
Agricultural/Unskilled labour	36
Cultivator	8
Fishermen	15
Skilled worker	5
Type of home	
Kachcha	58
Semi-pucca	24
Pucca	18
Own livestock	40
Own Agricultural land	17
Total N	3232

Media Exposure

An understanding of the media habits of the people becomes crucial as this provides inputs for deciding about the channel to be used for disseminating cyclone early warning messages. Table 10 gives the analysis of this data. TV emerges as the most popular channel of mass media, followed by radio and newspaper. The gender differentials are less pronounced in case of TV and radio, compared to the newspaper.

Table 10

Media Habits of People (past one week)

% Reporting	*Male*	*Female*	*All*
Ever read newspaper	56	19	35
Ever listened to radio	46	38	42
Ever watched TV	71	67	69

Cyclone Early Warning Messages

It would be most relevant to look into the aspects relating to cyclone early warning system in terms of source, advance duration, and response. As Table 11 indicates, TV and radio emerge as the most mentioned sources of cyclone early warning managed dissemination. Role of panchayat/government officials also gets highlighted in dissemination of information.

Table 11

Cyclone Early Warning Message System

Source of Information	%
Television	72
Radio	70
Dandora	43
Panchayat president	23
Government officials	23
Mike announcement	18
Advance Duration (Days)	
One	42
Two	43
Three +	13
Hours	
1-5	3
6-23	4
Response	
Meet and discuss about movement	31
Rush to shelters/High level places	27

Contd...

Contd...

Go to safe places	17
Go to neighbouring villages	12
Send children and aged people to safe places	4
Don't go anywhere	15
Response (night time)	
Go to strong buildings	34
Go to shelter	15
Go to school/college	9
Go to high level places	4
Remain at home	41
Do nothing, leave it to God	7
Total	3232

The duration of advance information varies from 1-5 hrs to 3 days. Thus, about half of the respondents reported that the cyclone early warning message reaches them one day or less in advance. This needs attention to facilitate action in terms of evacuation.

Majority of the respondents (85%) reported positive response to the messages but in case the message reaches them in the night, almost half of the respondents reported that they do not go anywhere and leave it to God. This again indicates the importance of timing of the message.

Contingency Plan

It is expected that all the vulnerable villages will have a contingency plan to mitigate the disaster – cyclone or flood. Availability of such a plan seems to be an issue due to the fact that only 4 per cent of the respondents reported availability of such plan. Not many of these respondents reported formation of the different committees, viz.

- Cyclone warning committee
- Shelter management committee
- Evacuation committee
- First aid committee
- Sanitation committee
- Relief committee
- Rehabilitation

Mitigation Facilities

Availability and adequacy of the disaster mitigation facilities in the villages covered can be understood from the results presented in Table 12. Though availability of temporary cyclone shelters is reported by most, more than one-fifth of the respondents reported non-availability of the same. Availability of country as well as motor boats needs attention. Adequacy of the facilities also emerges as an issue.

Table 12

Availability and Adequacy of Mitigation Facilities

Facility	*% Reporting*	
	Availability	*Adequacy*
Temporary cyclone shelter	91	47
Permanent cyclone shelter	78	37
Country boats	31	23
Motor boats	20	13

Concluding Remarks

The 9 coastal districts of Andhra Pradesh have experienced severe damage including loss of lives and property due to cyclones.

There is a need to improve the cyclone early warning systems through increasing the frequency of messages, making it well in advance so as to ensure evacuation of the families from the low-lying areas. The messages can be disseminated through TV and radio and mike announcements, and dandora as well at the village level.

There is an urgent need to prepare contingency plan in the coastal villages – particularly the vulnerable villages – followed by formation of the different committees to ensure community participation. NGOs and other community based organizations can be involved in this task.

The contingency plan should also address the specific needs of women, children, elderly and the physically challenged people.

The mitigation facilities need to be provided in accordance with the need and maintenance of the same needs to be reviewed from time to time, so that they can be used when needed.

Select References

Andhra Pradesh Voluntary Organizations and Social Worker's Association (1978): "Community Cyclone Shelters", in *Devastation and Development.*

Government of Andhra Pradesh (1979): Note on Recent Cyclone of May 1979.

– (1984): "Memorandum on Cyclone and Heavy Rains in AP", November 1984.

– (1987): "Memorandum on Cyclone and Heavy Rains in AP", November 1987.

– (1994): "Memorandum on Cyclone and Heavy Rains in AP", October–November 1994.

Reddy, A.V.S., Vinod Sharma and C. Mohana (2000): "Cyclones in Andhra Pradesh". *Oxfam, Hivox*, ECHO.

Stephen P. Cohen, and Raghavulu, C.V. (1979): *The Andhra Cyclone of 1977.*

12 Self-Reported Health Problems of Slum Dwellers in Pune

Anjali Radkar

Urban population is growing rapidly throughout the world. People are migrating to cities in search of work. The most unwanted consequence of urbanization is increase in slum population having large number settling down on relatively smaller land. Thus, slums increase the burden of already scarce resources and on overall urban infrastructure. Their living – large population residing in a smaller area with no or very limited amenities is undesirable. Environmental pollution in the urban areas is on the rise and all urbanites have no choice but to suffer from its consequences. Incidences of infectious diseases such as diarrhoea, malaria and respiratory infections remain high in poor dense urban settlements. At the same time, the incidence of heart disease, cancers, neurological diseases and other conditions associated with chronic exposure to toxic pollutants is also increasing driven by growth in industry, small-scale enterprise and traffic. Slum living along with pollution ill effects make these people more vulnerable in terms of health.

Dictionary meaning of the word 'slum' is "an overcrowded and squalid back street, district, etc., usually in a city and inhabited by very poor people". According to Slum Areas Improvement and Clearance Act, 1956, slum is defined as "any predominantly residential area where the dwellings which by reasons of dilapidation, overcrowding, faulty arrangement of design, lack of ventilation, light or sanitary facilities or any combination of these factors detrimental to safety, health or morals".

Slum settlements generally look similar with same degree of filth and squalor, poverty and malnourishment, disease and wretchedness. The usual sight is garbage dumps, choked gutters, malnourished children, queues on water taps, smoke filled atmosphere after cooking fires are lit. Such unhygienic atmosphere affects the health of the residents (Bapat, 1982).

Slum Scenario – Pune City

As in all major cities slum population in Pune is increasing and increasing fast. To understand what has happened to the slum population over the period, Table 1 is prepared.

Table 1

Slum Population within Pune Municipal Corporation

Year	*Slum Population (%)*
1951	7.6
1961	15.4
1971	28.0
1981	31.3
1991	33.6
2001	38.9

Source: Environmental Status Report, 2002–2003, Pune Municipal Corporation.

From Table 1 one can see the growth in the slum population of the city – in Pune Municipal Corporation (PMC) area. There is a large leap during 1961-1971 that is mainly because of the growth of the industry around Pune during that decade. Many people migrated in search of work then. During this and the decade followed by it, migration to cities was because of the series of droughts in Maharashtra. According to slum department of PMC, in 2001, about 39 per cent of the population of the city lives in slums. Such a large proportion of slum population also adds to already degrading environment of the city.

In PMC, 503 slum areas are identified. Out of these 503 slums 353 slums are declared slums and 150 are undeclared. To change the living conditions in the slums it is necessary to declare them. Only when slums are declared under Slums Act, local government can enter there and provide basic amenities – water, road, electricity and gutters – to slum dwellers. Among the slums 70 per cent slums are declared and hence are benefited by amenities provided by PMC.

Rationale of the Study

It is necessary to understand the impact of slum environment on the living of the residents as also on the health. However, it is difficult to measure the morbidity in population. The associated problem in the study of morbidity is the way of collecting information because reporting morbidity depends heavily on the perception of respondents. There is also a possibility that respondent though is aware of the problem may not reveal it. Still there is a need to have some information on morbidity at a community level.

Slum population bears dual burden of diseases – diseases due to life style, urban living and diseases due to poor economic status and thus, living conditions.

What really are the health problems in the slum areas is necessary when we need to understand more about this underprivileged population. Thus, an effort has been made in this cross sectional study irrespective of various limitations associated with the study like this.

Objectives

The objective of the paper is to portray the health status of slum dwellers of Pune as reported by themselves. The specific objectives would be:

- To understand the profile of slum population
- To assess the current morbidity as well as morbidity in one month reference period in Pune slums
- To understand and measure the chronic health problems among slum dwellers

Data

The data for the study comes from the survey conducted to assess the overall health status of the slum dwellers of Pune. The latest estimate (2003) of the slum population in PMC area was 11.25 lakhs. It was decided to study at least half a percentage of slum population or population of about 5,500 meaning visiting roughly 1,100 households.

It was decided that sample would be drawn from declared slums because in that case some basic information like the size of slum would be available. Thus 55 slums are selected from 353 declared slums from PMC using probability proportional to size sampling. From all the selected slums, 20 households are selected using systematic sampling technique from every slum to get the entire spread of the slum. Number 20 was fixed for every slum considering the fact that intra-slum variation in the pattern of health problems would be less compared to inter-slum variation. In actual study, data is collected from 1,095 households from 55 slums. In one slum, data collection had to stop because of societal resistance.

It was thought here that health conditions of the slum residents would be assessed on the basis of type of house, number of rooms, crowding in the house, bathroom inside or outside of the house and toilet facility.

Type of house is 'pucca' if all three – floor, walls and roof – are made using pucca material, is semi-pucca if any two of them are pucca material, semi-'kachcha' if one of them is pucca and kachcha if none of them is pucca.

Crowding in the house is computed as number of persons per room in the house.

The data collection of the study is carried out in January, which is a comparatively healthy period of the year than periods like rainy season or hot summer.

To understand the health problems information is analysed on the health problems faced in past one month as also the chronic health problems of the slum dwellers. It is possible that one person may suffer from multiple problems.

Information on health problems in past one month is collected by asking them what the problem was. Here the perception of the respondent plays a key role.

For chronic health problems broad classification is for illnesses namely, diabetes, cancer, heart diseases, arthritis, asthma, hypertension, acidity, stroke and gynaecological problems. Out of these arthritis can be occupation related – kind of physical activity. Diabetes, heart problems, hypertension, stroke are related to lifestyle and asthma and gynaecological problems are related to living conditions.

For chronic diseases the information is collected by directly asking them the name of the disease.

Results and Discussion

Profile of Pune Slums

Information about the slum households is presented in Table 2.

Table 2
Percentage of Households by Different Characteristics

Characteristics of Households	%
Type of house	
Pucca	25.6
Semi-pucca	52.0
Semi-kachcha	18.2
Kachcha	4.3
Ownership of house	
Owned	82.7
Rented	17.3
Separate kitchen	
Yes	42.4
No	57.6
Bathroom facility	
Inside the house	85.9
Outside the house	14.1
Toilet facility	
Independent	11.9
Shared	88.1
Source of water	
Separate tap	75.8
Common tap	24.2
Place of origin	
Pune	43.8
Place in Maharashtra	41.1
Place outside Maharashtra	15.1
Total	1095 = 100.0

Among 55 slums population covered is 6,131. Average family size of the slums in Pune city is 5.6. All those who reside in slums definitely belong to lower economic strata. The dwellings of slum residents are constructed using locally available materials and are generally of semi-permanent nature. Quarter of the houses are pucca houses, 52 per cent are semi-pucca, 18 per cent semi-kuchcha and less than 5 per cent are kachcha houses. Eighty-three per cent houses are owned by respondents and the rest 17 per cent are rented ones.

It is necessary to look at the conditions of the selected houses and facilities available to them. Dampness is a major determinant of ill health. Bapat (1982) established that child mortality is more in case of damp houses. In this study in 43 per cent houses, interviewers could not observe the dampness in the house, as they could not enter the house for collecting information. Because of the scarcity of space women responded standing at the door of the house. Dampness is observed in 15 per cent houses. Ventilation is poor in most of the houses.

About half of the houses are one-room houses. Separate kitchen is observed in 42 per cent of houses. In 86 per cent houses bathroom was inside the house. Availability of bathroom inside the house indicates two effects. In an already small house, bathroom would occupy the space and also would create dampness in the house. The other aspect of it is the house is big enough to include the bathroom inside giving the residents the required privacy and facility.

Toilet facility is important indicator of the living condition. Independent toilets are reported in case of 12 per cent houses and remaining 88 per cent are shared ones. Whether the children in the house and around use the toilet or not has a major impact on health of the residents. If the toilet is owned then there is all the possibility that children may use it. Owned toilets are maintained properly and hence increase the use. Shared toilets are comparatively unclean, faced with the problems – especially availability of water and thus could be used less by children.

Source of water for all the selected slums is piped water supplied by PMC. Among the selected households 76 per cent have the separate tap, whereas other 24 per cent get the water form a common tap provided by PMC. Electricity is almost universal in declared slums, having in 95 per cent of the selected households.

Apart from household characteristics it is also necessary to present personal characteristics of slum residents in Pune.

Sex ratio of the selected slums is 948. Age distribution shows 29 per cent of population below 14 years of age indicating high fertility there. Percentage of aged in the population is 5.5. It is less than the general population. Working class population is more in the slums because slum residence is closely related to migration and migration is mainly for the work. Twenty per cent of slum population is illiterate and 29 per cent is SSC or above. Ten per cent males and 24 per cent females are illiterate. For this study occupations of the slum residents are coded in detail, showing majority of them in the lower level occupations with unskilled work.

Table 3

Distribution of Slum Residents by Background Characteristics

Background Characteristics of Residents	*Per Cent*
Sex	
Male	51.3
Female	48.7
Age group	
0-4	8.6
5-14	20.6
15-59	65.3
60 or more	5.5
Educational level	
Illiterate	19.5
1-4 std	16.9
5-7 std	20.2
8-9 std	14.2
SSC	17.1
HSC	7.0
Diploma	1.1
Degree	4.0
Total	100.0 = 6131

Health Problems

Environmental factors affecting health of the slum population include crowding and congestion, unsanitary surroundings, lack of proper sunshine and fresh air. The health is also affected by communicable diseases of digestive system such as diarrhoea as also of respiratory system like bronchitis and pneumonia.

As stated before, information on current health problems as well as on health problems during past one month are recorded by asking the slum dwellers what did they suffer from during this period. The information is collected for all possible problems, however minor.

These are the self-reported problems sometimes in the form of symptoms and sometimes are diagnosed problems. The interviewers for this study are postgraduate students of health sciences who know about human physiology, anatomy as well as various communicable and non-communicable diseases, which increase the authenticity of the data. An attempt has been made here to record and classify even the minor complaint. These health problems are then sub-grouped in 29 broad categories that are presented in Table 4.

The data shows that among 6,131 slum residents 710 persons have reported at least one health problem in one month prior to study and 975 reported some problems they are suffering from currently. It gives the prevalence rate as 116 for health problems in one month and 159 for current illnesses. More people reported current problem compared to those who reported in a month before the survey. It may be because of recall lapse or the tendency to forget reporting the problem once it gets cured.

Current health problems reported by 975 slum persons are 1,337 and problems in one month reference period to 710 persons are 1,002.

It is seen that the most common health problem is cough and cold – 46 per cent of the total problems in past month and 44 per cent of the current. Flu, chills and fever are 9 per cent of the currently reported problems as against 14 per cent in a month before. This large proportion suffering from cough, cold and fever is explained by changes in temperature during winter, dry heat in the afternoon and cold at night. These changes affect relative humidity, which gives breeding ground to virus causing cold. Cold weather induces dryness in the lining of the nasal passage thus making person more susceptible to viral infection. Cold also leads to secondary bacterial infection of middle ear and sinus, causing fever. Slum population is under-protected by way of housing, for all these weather changes and more prone to bacterial infection.

Aches and pains are more here – 11 per cent in past month and 17 per cent currently – probably because the physical activity involved in their occupations. It was discussed before that slum residents generally have lower level occupations involving manual work. Such pains are

Table 4

Broad Classification of Current Health Problems and Health Problems in One-Month Period to Slum Residents of Pune

	Category	*Number of persons reporting current health problem **	*Number of persons reporting health problem in one month period***
1.	Dysentery, diarrhoea	9	19
2.	Worms, hepatitis, typhoid	2	10
3.	Herpes, chicken pox, measles	5	1
4.	Tuberculosis	1	1
5.	Dengue	1	1
6.	Flu, chills/fever	122	145
7.	Mental disturbance	2	2
8.	Weakness, loss of appetite, fatigue, weight loss	25	33
9.	Epilepsy, convulsions	6	6
10.	Eye problems	30	10
11.	Hearing problems	13	7
12.	Blood pressure	4	3
13.	Heart problem	21	13
14.	Cold and cough	590	464
15.	Dental problems	28	24
16.	Digestion related problems	76	44
17.	Acidity	11	7
18.	Skin problems	12	8
19.	Aches and pains	227	110
20.	Neurological problems	2	2
21.	Kidney, urinary problems	53	6
22.	Gyneacological problems	4	1
23.	Headache, migraine	64	39
24.	Accident	12	16
25.	Allergy	6	5
26.	Lump related/tumor	4	7
27.	Diabetes	1	-
28.	Pneumonia, asthma	2	5
29.	Hyperthyroidism	4	2

* Total may not add to 975 because of multiple problems.

** Total may not add to 710 because of multiple problems.

also because of housing conditions, dampness in the house. Headache forms 4 to 5 per cent of the problems. It also could be because of cold or fever related problems.

Different types of digestive problems form 4 per cent for one-month reference period whereas they are 8 per cent of the current problems. Dysentery and diarrhoea are recorded separately because they are related to unhygienic conditions in the slums and also unhygienic practices of cooking and storing food and water. Same is true for worms, jaundice and typhoid. Together they form 2 per cent of the past problems and 1 per cent of the current.

Weakness, loss of appetite, fatigue, weight loss form 3 per cent of the problems in past one month and 2 per cent of the current. These are caused by some other health problems also but one can also relate it to the poor nutritional status of the slum dwellers.

Seven persons have reported incidence of tuberculosis, 1 of dengue and 1 of chicken pox one month prior to study, whereas among current ones the numbers for these diseases are 1, 1 and 5.

Among the current health problems, 4 per cent of the problems are related to kidney or are urinary problems, which are significantly less (half a per cent) during one month reference period. Similarly problems related to eyes are also reported more as current problems than during the reference period. These problems may not be serious to report afterwards but they do disturb the person while suffering.

All these problems are as reported by the respondent spontaneously when asked about health problems in a month. To understand about some more problems, which could be left out, certain more probing questions are asked to get information on two stigmaticized diseases, tuberculosis and leprosy and also about two less reported, night blindness and dental problems.

Table 5

Symptoms of Tuberculosis

Symptom	*Yes*
Cough for more than 4 weeks	58
Fever in the evening	34
Expectoration	31
Chest pain	44
Weakness, weight loss, fatigue	56

For tuberculosis information is collected on cough for more than 4 weeks, fever in the evening, expectoration with or without blood, chest pain as also weight loss, weakness and fatigue.

When all these symptoms exist together, then one can see the possibility of tuberculosis infection. However, 1 person has already reported that they suffer from it. 12 persons report all the symptoms positively. These are cases of tuberculosis.

Exactly in the same fashion, information is collected on the symptoms of leprosy – whether have any light colour patches on the skin, itching and loss of sensation. From this information it is suspected that 5 persons suffer from it.

When asked directly about any skin problems, 80 persons responded positively. Changes in environmental conditions is a known cause of skin problems along with the causes associated with uncleanliness. Similarly, night blindness is reported by 17 persons, which is caused by vitamin A deficiency, indicating poor nutritional status. It is documented that dental problems persist in poor societies. In past one month 24 persons have reported the dental problems. However, when direct question is asked, the number shoots up to 583. The problems can be minor in some cases but are reported means the persons are suffering from it.

Chronic Health Problems

In the previous section, information of the health problems is collected on symptoms or actual diagnosed problem depending on the response. Here an attempt is made to know more on the health problems, which may not have been mentioned in the section before because once treated they do not cause day-to-day problem like diabetes and hypertension. There are also certain health problems like gynaecological problems, where women do not perceive it as a problem, hence tend to not to report it. Such few chronic problems are identified. While collecting the information directly it is asked that whether anybody in the family suffers from these problems. As stated before the list includes diabetes, cancer, heart disease, arthritis, asthma, acidity, stroke, blood pressure and gynaecological problems.

Eight hundred and seventy persons report at least one chronic health problem. Total prevalence rate of chronic health problems is 146 per 1,000 persons. It has come out clearly here that acidity is the most reported chronic health problem. Once again it is difficult to get the reasoning behind acidity. All other problems also are typically urban health issues. Slum population bears the dual burden of diseases, as diseases of the urban lifestyle as well as those with poor living. Diabetes, blood pressure and heart diseases are typically lifestyle related problems, which exist even in the slums irrespective of their comparatively more physical activity.

Table 6
Chronic Health Problems in Pune Slums

Chronic Health Problem	*Number **	*Prevalence rate per 1000*
Diabetes	108	18
Blood pressure	199	32
Cancer	5	1
Heart disease	33	5
Arthritis	127	21
Asthma	43	7
Acidity	247	40
Stroke	11	2
Gynaecological problems	151	25
Total	870	1000 = 6131

* Total may not add to 870 because of multiple problems.

Exposure to pollutants in atmosphere can give rise to respiratory problems as well as heart diseases. Slum dwellers are exposed more to the pollution because of their living conditions, having no protection from sun and being in areas where no tree cover exists.

Gynaecological problems include all problems and not just the infections, which are once again related to living conditions of slum dwellers (Radkar, 2005). Chronic problems are reported more for women because it includes gynaecological problems and arthritis, which also is predominantly female problem. Arthritis is related to heavy physical activity as well as housing conditions, dampness in the house.

Conclusions

A community study to assess the health status of slum residents of Pune city has been conducted. This paper sheds light on the health problems in one month prior to survey as well as of chronic health problems. For certain diseases an attempt has been made to present prevalence rates. Comparative figures for slum areas are not available as generally information is provided for rural and urban. But considering the conditions of slums and of slum dwellers separate figures need to be presented.

In this study from 1,095 households in 55 declared slums population covered is 6,131 with sex ratio 948. Twenty per cent are illiterates, 24 per cent females and 10 per cent males. Only about a quarter of the selected household live in a poorly constructed houses, a quarter in pucca houses and half of them in semi-pucca houses.

In slum areas cough and cold followed by flu, fever and chills are the major problems. After that aches and pains, digestive problems, diarrhoea and dysentery are the major ones. There are also infections like tuberculosis, dengue and chicken pox mentioned, though very less. Slum residence explains the extent of these problems.

Problems that are not reported spontaneously, but there was a doubt that such problems would exist, deliberate questions were asked to get 12 probable cases of tuberculosis, five cases of leprosy, 17 cases of night blindness and 583 dental problem sufferers. Among the chronic disease sufferers most are suffering from acidity, blood pressure and gynaecological problems and arthritis.

References

Bapat, Meera (1982): *Determinants of Health in Low Income Communities: A Case Study of Poona, Monograph*, published by Centre for Development Activities and Studies, Pune Environmental Status Report, (2002-2003): Pune Municipal Corporation.

Institute of Health Management Pachod, Pune (1998): *Census of Slums*, Pune.

Radkar, Anjali (2005): "Reproductive Health Problems of Slum Women in Mumbai, India: Slum – Non-slum Scenario", Unpublished paper.

IV Household Environment and Health

13 Health During Infancy and Early Childhood in India: Household Environment Matters

Usha Ram, Avishek Hazra and Sandip Chakraborty

The word 'environment' refers to whatever surrounds an object or any other such entity. Human populations experience environment in which they live as an assemblage of physical, chemical, biological, social and economic conditions which vary according to the local geography, infrastructure, season, time of day, and activities undertaken. The household environment carries the biggest risks to the health of people at large and to those poor segments of the population, including women and children, living in the developing regions. These risks are typically 'traditional' in nature, i.e., they often are associated with a lack of development (Gopalan and Saksena, 1999).

Polluted indoor air, contaminated water, lack of adequate sanitation, toxic hazards, disease vectors, ultraviolet radiation, and degraded ecosystems are all important and additional environmental risk factors during childhood. Particularly in developing countries, environmental hazards and indoor pollution are major contributors to childhood deaths, illnesses and disability caused by acute respiratory disease, diarrhoea diseases. Diarrhoea kills an estimated 1.6 million children each year globally, mainly due to drinking unsafe water and poor sanitation facilities. Indoor air pollution associated with the still-widespread use of biomass fuels kills nearly one million children annually, mostly as a result of acute respiratory infections (WHO, 2002).

In India, an estimated 350 million children below the age of 14 years live (accounting for nearly 30 per cent of the country's total population). It may be said that nearly 46 per cent of these children (about 158 million) belong to the age group of 0-6 years. This segment of the population, however, is unevenly spread over the country's 28 states and 7 Union Territories. Six states, viz., Uttar Pradesh, Bihar, Madhya Pradesh, Maharashtra, Rajasthan and West Bengal, account for about half of the child population (aged 0-6 years) of the country. Environmental factors play an important role as far as the mortality rates among the children aged under-five years of age are concerned. Further deaths due to diarrhoea under one year found to be more in developing countries, caused by 'exogenous' factors.

In the recent past, rapid deterioration in the quality of environment has overburdened the health system in many countries. Air and water pollution are directly responsible for many of the health problems and it is evident that the domestic (household) environment is a major contributor to this burden. The World Bank (1993(a,b)) estimated total burden of disease attributable to the problems related to the household environment including overcrowding, lack of facilities such as sanitation, garbage disposal, indoor air pollution and vector-breeding grounds, etc. The study further discussed that 30 per cent of the burden of disease could be avoided by improvements in the household environment, of which 20 per cent can be achieved through modest interventions alone. The most recent estimates of the WHO (1997) indicates that although majority of all cases have an environmental cause, 90 per cent of the diarrhoeal diseases could be avoided through 'feasible environmental interventions'. In all, environmental factors account for about 23 per cent of the global burden of disease.

Water and sanitation, are the two important factors for health and welfare of the society, where water quality, supply and hygiene play a major role. Supply of adequate quantity of water is a prerequisite for facilitating domestic activities and better hygiene practices. Water quality is linked to the health status of consumers. It is estimated that as many as 80 per cent of diseases in the world are associated with water usage or poor environmental hygiene. The infections can also spread through mosquitoes or insects through stagnated water and poor sanitation. Several studies in India have pointed out that there exist large variation within the country across states in all areas of development, be it socio-economic or demographic development or for that matter availability of the basic civic household amenities such as provision of safe drinking water, electricity, toilet facility, modern means of cooking to the households, housing adequacy, etc. (Ram and Annamma, 1998).

The recently launched National Rural Health Mission (NRHM) and the National Population Policy (NPP) 2000 have emphatically recognized the importance of health in the process of socio-economic development and improving the quality of life of the people of India. The NRHM goals and objectives seek to provide effective health care to the rural population, especially to the disadvantaged groups including women and children to begin with the 18 demographically backward states of India. The mission clearly underlines the need for integrating ongoing vertical programme of Health and Family Welfare and the synergistic approach of addressing not only health but also its determinants like sanitation, safe drinking water and nutrition. Research studies discussed above revealed that it is important to understand inter- linkages between health of the population particularly during infancy and early childhood and the household environment in which people live. The present paper thus attempts to do the following:

1. To discuss the existing scenario of household environment and health during infancy and childhood and changes in them over a period of three decades in the major states of India.
2. To examine the association between health during infancy and childhood and a few household variables.

Data and Methodology

The present study uses data mainly from three sources: Census, Sample Registration System (SRS) and National Family Health Survey.

The heath status during infancy and early childhood has been measured in the following ways:

1. Infant mortality rates from the SRS.
2. Mortality rates among children aged 1-4 years of ages by converting the probability of dying among children aged 1-4 years ($_4q_1$) from the SRS based abridged Life Table.
3. Proportion of children under age four (for the period 1992-93) and under age three (for the period 1998-99) suffering from Acute Respiratory Infection (ARI) and/or diarrhoea.

The household environment has been measured using data for basic household amenities from the household tables of the respective Censuses of India for the period of 1981, 1991 and 2001 with the help of following variables, percentage of households:

1. Having electricity.
2. Having toilet facility.
3. Using modern means of cooking (liquefied petroleum gas, electricity and biogas).
4. Having safe drinking water facility (tap water, hand pump and tube-well).
5. With two or fewer members per living room.

Simple correlation has been obtained between two sets of variables to understand the association. The analysis is carried out for 14 major states of India for rural and urban areas for the period 1981 through 2001. Thus data on infant and childhood mortality and household environmental condition refers to particular state and time period. This areal data analysis will throw light on variation in dependent variable by independent variables for different time periods.

As mentioned before, in the present analysis we attempt to understand the health status of the infants and very young children with the help of selected indicators of mortality and morbidity in this segment of the population. In this subsequent section we discuss each of them in details for India and its states.

Infant and Child Mortality

In the developed countries the levels of mortality during infancy seem to have reached the 'biogenetical minimum' as only a few newborn die (Gjaltema, 2001). However, in many developing countries like India, levels of infant and childhood mortality continue to be on the rather higher side. The continuing higher levels tolling in substantial deaths during the early childhood ages. As may be noted from the infant and child mortality rates (given in Appendix I) that though the mortality levels in the country have declined quite fast over a period of three decades, they continue to remain unacceptably high. In the beginning of the eighties, for every one thousand live births, about 110 infants died before celebrating their first birthday.

At the beginning of the present century, the number of infant death reduced to about 66 per one thousand, indicating a reduction of about 44 infant deaths (that is by one and a half unit annually) in a span of 30 years. Nevertheless, there are pockets of the country, particularly comprising of states in the northern regions, which failed to attain the kind of success observed at the national level. The rural-urban differentials in this context are quite visible. As may be seen, the mortality levels are typically higher in the rural areas as compared to the urban areas in all states. However, in most cases the rural-urban gaps are narrowing down with the passage of time.

The mortality rates among children aged 1-5 years have also declined over time. At the national level, the level of Child Mortality Rate (CMR) has reduced to 32 in 2001 from 41 and 71 during 1991 and 1981 respectively. It may said that the four backward states of Bihar, Madhya Pradesh, Uttar Pradesh and Rajasthan are contributing significantly to the persistent higher child mortality rate at the national level.

Childhood Morbidity

Prevalence of ARI among Children

In this section we will discuss the prevalence of two most important environment-related morbidity, viz., Acute Respiratory Infection (ARI) and diarrhoea. Among infants and children, acute respiratory infection, preliminarily pneumonia, is a major cause of illness and the leading cause of childhood mortality throughout the world (Murray and Lopez, 1996). As mentioned before, the data on ARI and diarrhoea has been taken from NFHS-I and NFHS-II reports. It may be reminded that the prevalence of ARI is based on the data for children born during the period 1-47 months for 1992-93 and children born during the period 1-35 months for 1998-99 prior to the survey. The results are given in Table 1. It may be clear that the values for the period 1998-99 is quite higher than that of 1992-93 and this is true for India and all the major states.

In 1998-99, around 19 per cent of children under age three who were ill with a cough accompanied by first breathing (which is a symptom of acute respiratory infection) during the past two weeks preceding the survey. The corresponding percentage for rural and urban areas are 20 and 16 respectively. In the states like Karnataka, Tamil Nadu, Gujarat, Haryana, Maharashtra and Punjab, children suffered from ARI varies in the relatively lower range of 8 to 14 per cent. In case of other states, the levels are relatively on the higher side. For example, about 20 to 24 per cent of the children suffered from ARI during the reference period in Bihar, Andhra Pradesh, Kerala,

Table 1

Prevalence of Acute Respiratory Infection (ARI) among Children Under Age 4 in 1992-93 and Children Under Age 3 in 1998-99 for India and Major States

India/States	*1992-93†*			*1998-99**		
	Combined	*Rural*	*Urban*	*Combined*	*Rural*	*Urban*
Proportion of children suffered from ARI						
Andhra Pradesh	6.5	6.9	5.1	19.3	20.3	16.2
Bihar	4.5	5.3	3.7	19.3	20.2	16.5
Gujarat	4.3	4.4	3.6	21.7	21.7	21.5
Haryana	5.8	6.8	3.7	11.0	11.4	10.4
Karnataka	5.4	5.6	4.5	11.8	11.9	11.6
Kerala	3.4	3.5	3.2	7.9	7.1	9.6
Madhya Pradesh	9.7	8.9	11.7	22.8	22.8	23.0
Maharashtra	4.7	5.1	3.4	29.2	30.9	23.4
Orissa	5.9	7.5	3.5	13.5	15.6	10.3
Punjab	10.4	10.9	7.7	22.5	22.4	23.3
Rajasthan	3.1	3.5	1.6	14.4	14.0	15.9
Tamil Nadu	4.9	5.6	1.2	22.0	21.3	24.5
Uttar Pradesh	8.6	9.6	6.8	10.3	10.0	10.7
West Bengal	7.2	7.2	7.0	21.1	21.5	18.9
Proportion of children suffered from Diarrhoea						
Andhra Pradesh	10.0	10.4	8.8	19.2	28.8	29.7
Bihar	11.7	11.6	12.1	15.0	16.1	14.7
Gujarat	13.7	14.3	9.6	17.7	15.6	17.9
Haryana	12.6	13.9	9.9	19.7	17.0	21.4
Karnataka	12.0	12.1	11.7	13.9	12.7	14.2
Kerala	9.7	10.4	8.1	13.9	12.4	14.6
Madhya Pradesh	9.2	9.9	7.3	11.6	12.7	11.3
Maharashtra	8.6	8.8	7.9	23.4	26.6	22.4
Orissa	9.7	10.2	8.8	25.4	29.1	22.9
Punjab	21.4	22.3	16.1	28.1	29.2	28.0

Contd...

Contd...

Rajasthan	11.0	12.1	7.2	9.8	11.0	9.4
Tamil Nadu	5.7	6.3	2.6	19.8	19.2	19.9
Uttar Pradesh	12.7	12.9	12.2	14.4	15.0	14.0
West Bengal	8.9	9.1	8.2	23.3	19.4	24.1

†Based on children born in the period 1-47 months prior to the survey.

*Includes surviving children of age 1-35 months from among the two most recent births in the three years preceding the survey.

Madhya Pradesh, Orissa, Rajasthan, Uttar Pradesh and West Bengal. The data apparently suggest that the situation seemed somewhat better during the early nineties. As at the national level, only 7 per cent of children under age four suffered from ARI symptoms in the two weeks before the survey and there is not much variation in the proportion for rural and urban areas. Only Orissa and West Bengal showing around 10 per cent of children suffered from ARI during that period. The interstate variation in this regard too is not remarkable. Thus, it may be said that the situation has deteriorated as far as the proportion of children suffered from ARI is concerned.

Prevalence of Diarrhoea

Diarrhoea is the second most important killer disease among children under age five worldwide, following acute respiratory infection. Deaths from acute diarrhoea are most often caused by dehydration due to loss of water and electrolytes. Table 1 gives the prevalence of diarrhoea among children under age 4 for NFHS-I and children under age 3 for NFHS-II.

The data in Table 1 suggest that about 10 per cent of children suffered from any type of diarrhoea during the two weeks before the survey and there is little difference in the prevalence of diarrhoea between rural and urban children. The proportion of children under age three suffered from diarrhoea in the two-week period before the interview increased to 19 per cent during 1998-99. No substantial difference is there in the prevalence of this morbidity between rural and urban areas. As may be noted from the data that the proportion of children who suffered from diarrhoea during the reference period is relatively higher in the backward states of Rajasthan, Madhya Pradesh,

Uttar Pradesh, Bihar and Orissa (18-23%). Contrary to this, the levels are on the lower side for the remaining states, more so states in the southern region.

Household Environment

In the present analysis we have used variables related to having electricity, safe drinking water, toilet facility, means of cooking and living rooms to understand the household environment that come into immediate contact with the child population. In this section we briefly discuss about them for India and its states.

Electricity

It may be observed from Table 2 that according to 2001 census, overall 56 per cent of the households in India live in the houses that have

Table 2

Percentages of Household Having the Facility of Safe Drinking Water and/or Electricity in India and Major States During 1981-2001

India/States	*1981*			*1991*			*2001*		
	Combined	*Rural*	*Urban*	*Combined*	*Rural*	*Urban*	*Combined*	*Rural*	*Urban*
Percentages of household having safe drinking water									
India	38.19	26.50	75.06	62.30	55.54	81.38	77.92	73.23	90.01
Andhra Pradesh	25.89	15.12	63.27	55.08	48.98	73.82	80.15	76.85	90.16
Bihar	37.64	33.77	65.36	58.76	56.55	73.39	86.59	86.11	91.23
Gujarat	52.41	36.16	86.78	69.78	60.04	87.23	84.09	76.87	95.40
Haryana	55.61	43.47	90.72	74.32	67.14	93.18	86.06	81.13	97.31
Karnataka	33.87	17.63	74.40	71.68	67.31	81.38	84.55	80.52	92.12
Kerala	12.20	6.26	39.72	18.89	12.22	38.68	23.39	16.88	42.84
Madhya Pradesh	20.17	8.09	66.65	53.41	45.56	79.45	68.43	61.51	88.55
Maharashtra	42.29	18.34	85.56	68.49	54.02	90.50	79.82	68.42	95.36
Orissa	14.58	9.47	51.33	39.07	35.32	62.83	64.19	62.88	72.32
Punjab	84.56	81.80	91.13	92.74	92.09	94.24	97.60	96.91	98.88

Contd...

Contd...

Rajasthan	27.14	13.00	78.65	58.96	50.62	86.51	68.18	60.45	93.52
Tamil Nadu	43.07	30.97	69.44	67.42	64.28	74.17	85.55	85.29	85.91
Uttar Pradesh	33.77	25.31	73.23	62.24	56.62	85.78	87.81	85.46	97.16
West Bengal	69.65	65.78	79.78	81.98	80.26	86.23	88.53	86.99	92.29
Percentages of household having electricity									
India	26.19	14.69	62.51	42.37	30.54	75.78	55.85	43.52	87.59
Andhra Pradesh	21.41	12.53	52.22	46.30	37.50	73.31	67.17	59.65	89.99
Bihar	9.20	3.48	50.09	12.57	5.57	58.77	10.25	5.13	59.28
Gujarat	44.81	30.83	74.40	65.93	56.43	82.96	80.41	72.12	93.39
Haryana	51.53	41.04	82.22	70.35	63.20	89.13	82.90	78.50	92.94
Karnataka	32.98	21.35	61.98	52.47	41.75	76.27	78.55	72.16	90.53
Kerala	28.78	23.21	54.57	48.43	41.95	67.65	70.24	65.53	84.34
Madhya Pradesh	17.11	6.89	56.42	43.30	34.49	72.52	69.98	62.32	92.26
Maharashtra	40.65	24.12	70.53	69.40	58.45	86.07	77.49	65.17	94.28
Orissa	17.75	13.03	51.74	23.54	17.45	62.11	26.91	19.35	74.08
Punjab	60.90	50.61	85.44	82.31	76.98	94.60	91.91	89.46	96.49
Rajasthan	20.54	8.70	63.67	35.03	22.44	76.67	54.69	44.02	89.61
Tamil Nadu	37.21	26.03	61.59	54.74	44.49	76.80	78.18	71.18	88.00
Uttar Pradesh	12.91	3.97	54.61	21.91	10.96	67.76	31.90	19.84	79.92
West Bengal	21.08	7.02	57.86	32.90	17.75	70.19	37.45	20.27	79.56

Source: Household Tables, Census of India 1981, Housing Tables, Census of India 1991 and 2001.

Note: Safe water is defined as source of drinking water either from tap or hand pump or tubewell.

electricity. The percentage of such households vary from a lowest range of 10 in Bihar to as high as 92 in Punjab. There are four states, viz., Orissa, Uttar Pradesh, West Bengal and Tamil Nadu in which the percentage of households living in electrified houses are lower than the national average. The percentage of households living in electrified houses in urban areas is invariably larger than that in the rural areas. This is true for all the states. The percentage of households living in electrified houses in rural areas varies from a low of 5 per cent in Bihar to well over 89 per cent in Punjab. The corresponding range for urban areas varies between 59 per cent in Bihar to 96 per cent in Punjab. In the previous decade, i.e., in 1991 the scenario was poor. The percentage of households living in electrified houses was only 42 at the national level comprising 31 per cent in rural areas and 76 per cent in urban areas. Almost the same picture existed that time too. Bihar is at the bottom with a percentage of 13 and as usual Punjab is on the top with a percentage of 85. If we go back more, the situation is more atrocious. Only 26 per cent of the households in India enjoyed the electricity facility. The percentage for the same was 15 in rural areas and 63 in urban areas.

Sanitation

At the outset it may be observed from Table 3 that as is the case at national level, at state level too the sanitary conditions are not very encouraging and said to be indeed miserable in rural parts of India in almost all the states except Kerala. As it is seen, only 36 per cent of the households in India have any toilet facility. The corresponding figures are 22 and 74 respectively for rural and urban areas. If we see the state-wise variation, it may be noticed that there exists a wide variation with respect to this amenity. On one side, there are states like Orissa, Bihar and Madhya Pradesh where just about 15 to 24 per cent of the households having toilet facility. On the other extreme, we have Kerala which stands out in this respect with well above 84 per cent of the households having this facility. Quite naturally the figure tells that the sanitary condition is far better in urban areas than their rural counterparts in all the states. It is needless to depict vividly the poor sanitary situation in 1991. The table shows that at the national level, only 24 per cent of the households were having toilet facility and the corresponding figures for rural and urban areas were 9 and 64. During that period the country experienced similar wide state-level variation. Information on toilet facilities was not collected for the rural areas in 1981 census. Data for distribution of households having toilet facilities are only available for urban areas. So the percentages of households calculated for urban areas have been put for the combined areas. It is evident that the situation of urban areas in terms of sanitation was far below than the satisfactory level. Only 58 per cent of households in urban India had the toilet facility.

Table 3

Percentages of Household Having Toilet Facility in India and Major States for 1981, 1991 and 2001

India/States	1981			1991			2001		
	Combi-ned	Rural	Urban	Combi-ned	Rural	Urban	Combi-ned	Rural	Urban
India	58.15	-	58.15	23.70	9.48	63.85	36.40	21.92	73.72
Andhra Pradesh	44.07	-	44.07	18.40	6.62	54.60	32.99	18.15	78.07
Bihar	52.95	-	52.95	11.75	4.96	56.54	19.19	13.91	69.69
Gujarat	60.11	-	60.11	30.69	11.16	65.71	44.60	21.65	80.55
Haryana	58.09	-	58.09	22.45	6.53	64.25	44.50	28.66	80.66
Karnataka	53.28	-	53.28	24.13	6.85	62.52	37.50	17.40	75.23
Kerala	59.14	-	59.14	51.28	44.07	72.66	84.01	81.33	92.02
Madhya Pradesh	52.73	-	52.73	15.07	3.64	53.00	23.99	8.94	67.74
Maharashtra	59.37	-	59.37	29.56	6.64	64.45	35.09	18.21	58.08
Orissa	41.88	-	41.88	9.81	3.58	49.27	14.89	7.71	59.69
Punjab	64.75	-	64.75	33.18	15.79	73.23	56.84	40.91	86.52
Rajasthan	56.48	-	56.48	19.57	6.65	62.27	29.00	14.61	76.11
Tamil Nadu	51.27	-	51.27	23.13	7.17	57.47	35.16	14.36	64.33
Uttar Pradesh	62.06	-	62.06	18.02	6.44	66.54	31.43	19.23	80.01
West Bengal	77.74	-	77.74	31.51	12.31	78.75	43.71	26.93	84.85

Source: Household Tables, Census of India 1981, Housing Tables, Census of India 1991 and 2001.

Note: Information on toilet facilities was not collected for the rural areas in 1981 census. Data for distribution of households having toilet facilities are only available for urban areas. So the percentages of households calculated for urban areas have been put for the combined areas also.

Fuel Used for Cooking

For this, we have computed proportion of households using modern means of cooking. The data on the type of fuel used for cooking presented in Table 4 reveals that a large proportion of the households in India depends on the traditional sources of fuel such as cow-dung cake, coal, charcoal, wood, kerosene, etc., for cooking and which are hazardous to family's health in general and to the women's and child's

Table 4

Percentages of Household Using Modern Means of Cooking in India and Major States for 1991 and 2001

India/States	*1991*			*2001*		
	Combined	*Rural*	*Urban*	*Combined*	*Rural*	*Urban*
India	8.74	1.81	28.33	18.18	6.28	48.94
Andhra Pradesh	8.34	2.62	25.91	19.76	10.84	46.89
Bihar	2.35	0.34	15.62	4.00	1.00	32.86
Gujarat	17.60	4.43	41.19	29.48	10.30	59.60
Haryana	13.40	2.23	42.74	30.92	15.99	65.07
Karnataka	9.38	1.84	26.12	20.04	6.35	45.82
Kerala	5.20	1.86	15.13	18.63	12.75	36.23
Madhya Pradesh	6.48	0.72	25.58	14.18	2.77	47.45
Maharashtra	17.73	4.90	37.26	30.55	10.77	57.61
Orissa	3.21	0.93	17.64	6.48	1.83	35.59
Punjab	16.67	4.60	44.46	34.45	19.01	63.23
Rajasthan	6.46	0.64	25.73	15.69	4.18	53.49
Tamil Nadu	8.21	2.03	21.51	19.64	7.00	37.38
Uttar Pradesh	5.94	0.97	26.76	11.69	2.87	46.88
West Bengal	5.30	0.84	16.29	13.02	2.72	38.52

Source: Household Tables, Census of India 1991 and 2001.

Note: Modern means of cooking is defined as type of fuel used for cooking is either LPG or electricity or biogas.

health in particular. Only 18 per cent of households have reported using LPG or electricity or biogas. The percentage of such households is virtually small in rural India, as they constitute around 6 per cent of the total households. The corresponding figure in urban areas is about 49 per cent. The percentage of households using modern means of cooking lies between a wide ranges of 4 in Bihar to 34 in Punjab. Looking at the scenario of 1991 census, it comes out that overall 9 per cent of households depended on modern means of cooking. The corresponding percentage in rural areas is negligible, less than two per cent and the same for urban areas is around 28 per cent. It is worthwhile to mention here that no information was collected regarding type of fuel used for cooking in 1981 census. It may be emphasized that less than 15 per cent of the households in the states of Bihar, Madhya Pradesh, Rajasthan and Uttar Pradesh reported use of modern means of cooking in 2001.

Safe Drinking Water

The data on percentages of households having access to safe drinking water bring out clearly that in 2001 around 78 per cent of the households in India had access to safe drinking water. The percentage of such households in urban and rural areas is about 73 and 90 respectively. In other words, about a quarter of the rural households and 10 per cent of the urban households in India did not have access to safe drinking water. Table 2 indicates that except Kerala, the situations of all the states are similar and around 80 to 90 per cent households having the facility of accessing safe drinking water. In Kerala, it may be specified, the source of drinking water from well is regarded as safe since most of the wells are covered and are located within the premises. For the present analysis this has not been taken into account to maintain the uniformity. Evidently, the percentage of households having access to safe drinking water is higher in urban areas in all the states. If we look at the scenario two decades back, the situation was not good. According to 1981 census, only 38 per cent households in India had access to safe drinking water. Though in urban areas 75 per cent households had the facility to access safe drinking water, the situation in rural areas was pathetic. Only 27 per cent of households had this facility. In 1991, the situation improved a bit. The table reveals that on an average around 62 per cent of households in India had access to safe drinking water. The percentage of such households in urban and rural areas is about 56 and 81 respectively. Looking seriously at these percentages over decades, one may observe that Punjab remains on the top in all the three time periods as the proportion of households having access to safe drinking water is highest. The situation in this respect is relatively poorer in the states of Madhya Pradesh, Rajasthan and Orissa.

Crowding in the House

In order to assess the crowding in the house across the states, we have calculated proportion of households with two or fewer persons per room. Table 5 provides data on the percentage of households with two

Table 5

Percentages of Household Having Facility Up to Two Persons Per Room in India and Major States for 1981–2001

India/States	*1981*			*1991*			*2001*		
	Combined	*Rural*	*Urban*	*Combined*	*Rural*	*Urban*	*Combined*	*Rural*	*Urban*
India	32.49	31.80	34.67	43.43	42.09	47.17	36.66	34.09	43.27
Andhra Pradesh	30.70	29.67	34.26	43.62	41.94	48.74	39.36	37.04	46.40
Bihar	33.21	32.74	36.55	41.09	40.44	45.37	28.98	28.43	34.23
Gujarat	24.46	21.11	31.57	36.67	30.92	46.97	31.06	25.66	39.51
Haryana	32.00	30.84	35.41	49.08	48.46	50.72	40.89	39.42	44.23
Karnataka	29.64	29.57	36.43	44.01	40.73	51.12	37.88	32.69	47.63
Kerala	44.79	44.08	48.07	69.79	69.20	71.57	65.41	64.35	68.61
Madhya Pradesh	34.22	32.64	40.32	47.58	45.57	54.24	35.77	32.41	45.54
Maharashtra	21.83	19.59	25.66	30.66	26.74	36.30	33.18	30.00	37.53
Orissa	38.26	37.81	41.39	50.89	50.41	53.95	40.56	39.63	46.34
Punjab	30.57	28.48	35.54	49.71	48.43	52.63	42.60	40.92	45.73
Rajasthan	32.97	31.40	36.10	44.76	42.67	51.66	34.49	31.97	42.77
Tamil Nadu	33.85	32.66	36.44	45.00	42.77	49.78	42.03	37.72	48.08
Uttar Pradesh	40.30	41.35	35.41	47.09	48.07	42.96	32.36	31.99	33.83
West Bengal	23.21	20.03	31.66	32.28	28.72	41.06	30.29	25.87	41.11

Source: Household Tables, Census of India 1981, Housing Tables, Census of India 1991 and 2001.

or less persons per room. It may be observed from the data that around 37 per cent of households live in the houses with two or fewer person per room (in other words, may be referred as those having adequate housing space). The proportion of such households is slightly higher in urban areas compared to the rural areas. The corresponding percentages for urban and rural areas are 43 and 34 respectively. Except Kerala, where around 65 per cent households seem to have sufficient housing space for their inmates. In all other states, percentage of such households are below 50 per cent. Looking to the situation of the previous two decades, it is visible that there is not much change over time. As it is seen, in 1981, at all India level the percentage of households having adequate housing space was 33 which increased 10 units and in 1991 the percentage of such households was 43. In 1981, the percentage of households having this facility in rural areas varies from a lowest of around 20 in Maharashtra and West Bengal each to 44 in Kerala. For urban areas the range is 26 in Maharashtra to 48 in Kerala. If we try to find out the proportion for the same in 1991 census, the data enlighten that in rural Kerala the percentage is quite satisfactory, 69 per cent of households have adequate housing space. Except Maharashtra, West Bengal and Gujarat where the percentage is around 30, in the remaining states 40 to 50 per cent households have adequate housing space. In 2001, around two-thirds or more households in Bihar, Madhya Pradesh, Rajasthan and Uttar Pradesh had on an average more than two persons living in one room.

Results of Correlation Analysis

We have calculated the correlation coefficients between the selected indicators for household environment and those related to health status during infancy and early childhood and these are presented in Table 6. Since NFHS-I took place in 1992-93 and NFHS-II took place in 1998-99, we have assumed that they would reflect the similar picture of 1991 and 2001 and would not hamper the analysis. It is clear from the Table 6 that in the early nineties, there is a strong negative association between most of the household environment indicators with both level of Infant Mortality Rate (IMR) and Child Mortality Rate (CMR). The correlation coefficient between IMR and percentage of households accessing safe drinking water is 0.75 (negative) and it is statistically significant. Similarly the correlation coefficient between IMR and percentage of households having electricity and percentage of households using modern means of cooking is 0.57 (negative) in both the cases. The correlation coefficients between CMR and the other household environment indicators (except that related to the overcrowding) lie between 0.59 (negative) to 0.69 (negative) and all are significant. To our surprise, the relationship between child morbidity and household environment variables is rather weak. For example, the association between diarrhoea and percentage of household accessing safe drinking water is very poor ($r = -0.17$).

It may be observed from Table 6 provides the values of 'r' for the same for the period circa 2001. As may be seen, safe drinking water, modern means of cooking and toilet facility each are significantly correlated with IMR and the relationship is negative. In other words, it means that the states, which have higher proportion of households

Table 6

Correlation Coefficients for the Selected Household Environment Indicators and Infant and Child Morbidity and Mortality Indicators, 1991-2001

	Per cent HH drinking safe water		*Per cent HH having electricity*		*Per cent HH using modern means of cooking*		*Per cent HH having toilet facility*		*Per cent HH with 2 or less members living in 1 or 2 rooms*	
	1991	*2001*	*1991*	*2001*	*1991*	*2001*	*1991*	*2001*	*1991*	*2001*
Combined										
IMR	-0.75**	-0.69**	-0.57*	-0.43	-0.57*	-0.55*	-0.70*	-0.64**	-0.42	0.02
CMR	-0.59*	-0.08	-0.69**	-0.69**	-0.56**	-0.70**	-0.69**	-0.45	0.35	-0.49
ARI	–	–	-0.38	-0.30	-0.37	-0.31	–	–	-0.16	-0.18
Diarrhoea	-0.17	-0.48	–	–	–	–	-0.16	-0.54*	–	–
Rural										
IMR	-0.69**	-0.67**	-0.50	-0.31	-0.55*	-0.47	-0.61**	-0.59*	0.41	0.15
CMR	-0.55*	-0.29	-0.55*	-0.70**	-0.60*	-0.73**	-0.46	-0.45	0.40	-0.47
ARI	–	–	-0.32	-0.62*	-0.18	-0.58*	–	–	-0.23	-0.22
Diarrhoea	-0.40	-0.75**	–	–	–	–	-0.32	-0.57*	–	–
Urban										
IMR	-0.39	-0.24	-0.50	-0.42	-0.25	0.12	-0.44	-0.02	0.30	-0.27
CMR	-0.05	-0.26	-0.49	-0.62*	-0.23	-0.38	-0.14	0.06	0.01	-0.50
ARI	–	–	-0.48	-0.51	-0.50	-0.43	–	–	-0.24	0.15
Diarrhoea	-0.48*	-0.40	–	–	–	–	-0.66*	-0.54*	–	–

Note: Kerala has been excluded from the analysis due to the absurdness in the value of safe water.

**Correlation is significant at the 0.01 level (2-tailed).

*Correlation is significant at the 0.05 level (2-tailed).

enjoying the basic civic amenities, are likely to experience lower level of IMR. Again, per cent households having electricity, using modern means of cooking, having toilet facility and adequacy in housing space each are negatively and significantly correlated with CMR. That means the states where larger proportions of households do not have these facilities are likely to experience higher child mortality. We have also tried to find out the association between ARI and percentage of households having electricity, using modern means of cooking, having adequate housing space. The results indicate that there exists negative correlation between each of these indicators and percentage of children suffered from ARI. Though the correlation coefficient is not statistically significant, there is negative association between percentage of children suffered from diarrhoea and percentage of households having safe drinking water facility. That means higher percentage of households accessing safe water, are less likely to suffer from the water-borne disease – diarrhoea. A strong negative relationship existing between percentage of children suffered from diarrhoea and percentage of households having toilet facility and that association is significant. Similar picture is depicted from the table for the rural areas as well.

Conclusion

The present paper discusses the current levels and past trends in infant and childhood mortality as well as morbidity situation, for India and major states along with various components of household environment. The paper also attempts to find out the association between indicators related to health status during infancy and early childhood and some selected indicators for household environment. The analysis is based on three consecutive census data of 1981, 1991 and 2001. The data for morbidity has been taken from NFHS-I and NFHS-II.

Many states are far from satisfactory level in terms of possessing basic household amenities. Simple correlation of IMR and CMR with selected household environment indicators signifies the importance of basic household amenities in improving the health status during infancy and early childhood. The availability of safe drinking water, toilet facilities and modern means of cooking facility seems to play a very crucial role in explaining the differentials in IMR and CMR. So, it may be concluded that provision of such basic amenities and basic health services in a community would go a long way not only to reduce the infant and child mortality but also reduce the state-wise variation in India.

Appendix I

Infant Mortality Rates in India and Major States for 1981-2001

India/States	*1981*			*1991*			*2001*		
	Combined	*Rural*	*Urban*	*Combined*	*Rural*	*Urban*	*Combined*	*Rural*	*Urban*
Infant mortality rates									
India	110	119	62	80	87	53	66	72	42
Andhra Pradesh	86	93	52	73	77	56	66	74	40
Bihar	118	124	60	69	71	46	62	63	52
Gujarat	116	123	89	69	73	57	60	68	42
Haryana	101	108	52	68	73	49	66	68	55
Karnataka	69	77	45	77	87	47	58	69	26
Kerala	37	40	24	16	17	16	11	12	9
Madhya Pradesh	142	152	80	117	125	74	86	92	53
Maharashtra	79	90	49	60	69	38	45	55	28
Orissa	135	140	68	124	129	71	91	94	61
Punjab	81	88	51	53	58	40	52	55	37
Rajasthan	108	118	53	79	84	50	80	84	57
Tamil Nadu	91	104	55	57	65	42	49	55	35
Uttar Pradesh	150	157	97	97	102	74	83	86	62
West Bengal	91	98	44	71	76	47	51	54	37
Mortality Rate among children aged 1-4 years									
India	71	79	37	41	46	21	32	35	18
Andhra Pradesh	53	59	28	24	27	12	13	16	9
Bihar	81	88	42	53	44	32	61	68	33
Gujarat	55	58	36	35	41	22	30	31	29
Haryana	45	47	22	31	35	18	23	27	6
Karnataka	55	65	28	28	32	16	16	19	7
Kerala	21	22	20	6	6	6	5	3	6
Madhya Pradesh	97	107	46	63	71	28	50	56	14

Contd...

Contd...

Maharashtra	50	58	31	22	24	17	11	13	6
Orissa	62	64	41	48	50	23	35	35	32
Punjab	42	46	27	23	25	17	11	10	17
Rajasthan	86	94	49	50	54	29	49	52	31
Tamil Nadu	63	74	36	21	25	12	49	16	17
Uttar Pradesh	108	116	55	60	65	35	83	42	31
West Bengal	52	59	26	27	30	17	51	41	19

Source: SRS for 1981, 1991 and 2001 and SRS based abridged Life Table for the period 1976-1980 and 1980-85, 1989-93 and 1998-2002

References

Bruce, N., R. Perez-Padilla, and R., Albalak (2000): "The Health Effects of Indoor Air Pollution in Developing Countries", *WHO Bulletin* 78: 1080-1092.

Gjaltema, Taeke A. (2001): *Modeling Infant and Child Mortality in India*, Population Research Centre Master thesis series, University of Groningen.

Gopalan, H.N.B and S. Saksena (1999): *Domestic Environment and Health of Women and Children*, Vedams Books (P) Ltd., New Delhi.

IIPS and ORC MACRO, "National Family Health Survey (NFHS-1), India 1992-93", Mumbai, International Institute for Population Sciences.

–, "National Family Health Survey (NFHS-2), India 1998-99", Mumbai, International Institute for Population Sciences.

Murray, Christopher J.L and Alan D. Lopez (1996): *The Global Burden of Disease*, Cambridge, Massachusetts: Harvard University Press.

Ram, Usha (1998): *Household Amenities in the Context of Development in Maharashtra: District Level Analysis of 1991 Census*, Gokhale Institute Mimeograph Series No. 49. Population Research Centre, Gokhale Institute of Politics and Economics, Pune.

Ram, Usha and K.I. Annamma (1998): *Levels and Differentials in Infant and Child Mortality in Maharashtra: District Level Analysis*, Indian Association of Social Science Institutions (IASSI), New Delhi 16 (3-4):74-105.

The World Bank (1993)a: "World Development Report. Investing in Health – World Development Indicators", New York, Oxford University Press,. pp. 329.

– (1993)b: "Water Resources Management: A World Bank Policy Paper", Washington DC, International Bank for Reconstruction and Development.

WHO (1972): "Health Hazards of the Human Environment", World Health Organization, Geneva.

– (1997): "Health and Environment in Sustainable Development: Five Years After the Earth Summit", World Health Organization, Geneva. p. 242.

– (2002): "The Environment and Health for Children and Their Mothers", World Health Organization, Geneva.

14 Housing Environment and Child Morbidity in Rural Andhra Pradesh and Tamil Nadu: An Analysis of the NFHS-2 Data

N. Audinarayana

Mortality rate for infants and children under age five has been decreasing over the years, although the rate of decline of morbidity has been rather slow in most of the developing countries. The attempts made by the investigators in quantifying, and comparing illness and disease between populations and changes over time have grown its importance as mortality declines to very low levels. Infants and young children up to the age of four years are by far the most vulnerable to health risks within the immediate family and community environment. The ability to survive the first few years of life and the quality of that survival is a function of many environmental and social stresses that impinge upon the individual child beginning during pregnancy and continuing through infancy and childhood. The common diseases that affect children are fever, cough and diarrhoeal diseases and respiratory ailments (acute respiratory infection, ARI).

In studying child mortality/morbidity, biomedical researchers give little attention to socio-economic determinants of disease except to note them as background variables and concentrates on the disease causing agents. On the contrary, social scientists seek socio-economic explanations for diseases. Both frequently tend to forget the common area, viz., the environment, which is determined by the socio-economic factors and which determines the disease causing agents. It is known that diseases occur due to imbalances caused between man and his environment. The agent and host factors have been identified and studied but the environmental factor of which the patient is a part remains an unknown territory. And, without the knowledge of the environment, it is difficult to prevent or control the diseases. Moreover, diseases and health are the direct outcome of a set of factors originating in the social conditions of life and behaviours of families. By and large, it is the combination of socio-economic and housing conditions in which families live, in addition to health facilities and services, which generally reduce the prevalence of diseases, determine the actual health outcomes. Identification of such factors, which may produce some differentials in morbidity pattern among children, will be helpful to policy makers in their work to reduce child morbidity.

With this background an attempt is made in this paper to examine whether housing environmental factors would have any influence on the three common child morbid conditions, viz., diarrhoea, fever and cough. Attempt is also made to examine such effects controlling for (in the presence of) selected socio-economic characteristics of parents. Further, these associations have been examined for the data collected from rural areas of Andhra Pradesh and Tamil Nadu, as the housing environmental factors would be more vulnerable in rural settings. Further, both these states share the same monsoon seasons compared to northern and western India. This assures a control over season, which is one of the important factors in determining the incidence of certain water borne and other communicable diseases.

Theoretical Considerations and Earlier Research

During childhood man is completely dependent for his survival on the environment in which he is born. Next to food and clothing, shelter is one of the most important essentials of life. A house defined as "the physical structure where a family lives provides shelter from the weather". Today, a house is not simply a place providing shelter to people or protecting them from heat, light and rain, but it is the part of the man's total environment and greatly responsible for the health and also for the occurrence of disease among the dwellers. The World Health Organization (WHO) considers the "residential environment" as "the physical structure that man uses and the environs of that structure including all necessary services, facilities, equipment and devices needed or desired for the physical and mental health and the social well-being of the family and the individual" (Park and Park, 2000).

It is widely argued that a variety of housing environmental factors would affect the child morbidity. In a study conducted among 893 children under five years of age in and around of Varanasi, Agarwal and Katiyar (1981) noticed that children who lived in pucca and mixed houses in urban slum and rural areas had significantly higher morbidity compared to the urban counterparts. The incidence of morbidity was higher among children who lived in adequately ventilated kachha houses, had poor source of water supply through open wells and practised open field defecation compared to those who lived in pucca houses with adequate ventilation, utilized tap water, and were using service latrines. In a study in Lagos, Chojnacka and Adegbola (1984) found that the five leading groups of diseases observed a higher frequency among children from households with both or one of the amenities than from households with none. According to the authors, such relationship could be that children from households without these basic amenities have acquired a higher immunity and thus the level of resistance particularly to infective diseases is better in general, than in case of children from household with basic sanitary facilities. On the other hand, higher room density in the child's household is positively associated with higher frequency of diseases of poor environment. Visaria (1988) reported that episodes of diarrhoea, and infections of the respiratory tract as well as malaria, etc., result partly from a contaminated water supply, inadequate faeces disposal and unhygienic conditions at home. According to UNICEF (1991), a major environmental problem for a large proportion of mankind is inadequate or unsafe water supply, which together with unsafe sanitation is responsible for perhaps three-quarters of all infections and most child deaths in the developing world. Based on the Demography Health Surveys, Boerma et al. (1991) found that in seven of the 21 countries the risk of diarrhoea is one-fifth higher among children having no radio in their households compared to those having radio. In 10 of the 21 countries, the risk of diarrhoea is more among children who have access to piped water supply in their households. In Bostwana and Kenya, the prevalence of diarrhoea is higher among children having toilet facilities compared to those with no such facility. Majumdar and Islam (1993) citing the work of Chen et al. on Matlab data concluded that inadequate sanitation is the major factor for diarrhoea among children under five years of age. Contaminated water source and other unhygienic practices in the household are the causes for the high incidence of various communicable or infectious diseases. Based on the National Family Health Survey, 1992-93 data, Mishra and Retherford (1997) noticed that ARI rate is almost one-third higher for children living in kachha houses than children living in pucca or semi-pucca houses. The rate is almost one-third higher for children living in households that use biomass fuels. Based on the NFHS-1 data for Andhra Pradesh and Tamil Nadu, Vidya (1997) reported that while children living in pucca houses have less chance of being affected by fever and cough than the rest, using surface water for non-drinking purpose tend to increase the odds of getting fever by children than those using piped/hand pump/well in residence.

Majority of the studies discussed earlier, by and large, showed that housing environment affects the prevalence of morbidity among children under 3 or 5 years of age. However, prior research and literature also demonstrated that socio-economic characteristics of the members (or parents) who live in the households, on the one side, would affect the housing environment; on the other side, they would have a direct effect on the morbidity conditions of their children (Chojnacka and Adegbola, 1984; Tekce and Shorter, 1984; Muttalib, 1990; Sethapongkul, 1992; Timaeus and Lush, 1995; Vidya, 1997; Sivakami, 2001).

Sources of Data and Methods

The data required for the present paper were drawn from the National Family Health Survey, 1998-99 for the states of Andhra Pradesh and Tamil Nadu (IIPS, Mumbai and ORC Macro, 2000; 2001). Information for the selected variables (as shown in the lines that follow) had been matched from household file and children's file, and then computed/recorded for analysis. In all, there were 829 and 748 children, respectively born during the last three years preceding the survey in rural areas of Andhra Pradesh and Tamil Nadu, of which for 50 and 32 children information about their morbidity conditions was not available and therefore, not considered for analysis. Thus, the final sample works out as 779 and 716 children for respective states.

The analysis is done at three stages. First, the prevalence rates of diarrhoea, fever and cough among the sample children have been computed based on the formula given in the following lines. At the second stage, the differentials in the prevalence rates of the three morbidities under consideration have been computed across the selected socio-economic characteristics of the parents and housing environment factors. Finally, the determinants of different child morbid conditions have been examined with the help of logistic regression analysis (Retherford and Choe, 1993).

Results and Discussion

Prevalence of Morbidity among Children

In the present context, period prevalence rates have been computed. Such rates would measure the frequency of all current cases (old and new) existing during a defined period of time expressed in relation to a defined population. It includes cases arising before but extending into or through the period as well as those cases arising during that period.

$$\text{Period Prevalence Rate (persons)} = \frac{\text{Number of persons who are sick some time during a definite period}}{\text{Average number of persons exposed to risk during that period}} * 100$$

Measuring morbidity based on data in two weeks of time and reported by mother has some limitations also. The important one is the seasonal variations of diseases. So, we cannot take that prevalence rate of an illness as reflecting what would be obtained if the entire year is considered. The two weeks reference period varied between individuals depending on the date of interview as the reference period is two weeks preceding the survey date, which varied substantially between individuals as reported earlier. This adds to the problem of reference period for the morbidity rate. The rate computed may therefore be combined to give a rough magnitude of the disease problem.

On the whole, the prevalence rates of diarrhoea, fever and cough, among the rural children and reference periods under consideration, observed to be comparatively higher in Andhra Pradesh than Tamil Nadu. Further, the prevalence rate of cough found to be the highest among the three illnesses considered here followed by fever, but the prevalence of diarrhoea is the least, irrespective of the states under consideration.

Table 1

Prevalence Rates of Diarrhoea, Fever and Cough among Rural Children Born during Three Years Preceding the NFHS, 1998-99 Across the States of Andhra Pradesh and Tamil Nadu

State/Disease	*Prevalence Rate*	*Frequency*
Andhra Pradesh (N = 779)		
Diarrhoea	14.5	113
Fever	27.9	217
Cough	35.9	280
Tamil Nadu (N = 716)		
Diarrhoea	13.8	99
Fever	21.4	153
Cough	26.0	186

Differentials in Child Morbidity Conditions

Differentials in Diarrhoea

In rural areas of Andhra Pradesh state, the prevalence rate of diarrhoea is observed to be more among children living in semi-pucca houses closely followed by kachha houses, whose drinking water source is well, river, stream, etc.; who do no have a toilet facility at their house, households in which kerosene is used for lighting and biomass fuels for cooking than their counterparts. Unexpectedly, children living in households with high density and which do not have a separate room for kitchen showed a lower prevalence of diarrhoea. Across the socio-economic variables, the prevalence of diarrhoea for children is more among those whose mothers are illiterate closely followed by middle school completed compared to those whose mothers have completed high school and above and literate, but middle school not completed. Conspicuously, diarrhoea is more prevalent among those children whose mothers are working and living in households that have less durable goods than their counterparts (col. 2 of Table 2).

In the case of rural Tamil Nadu, contrary to the expectation, diarrhoea is more prevalent among children living in pucca houses, having some type of toilet facility and using kerosene/gas/electricity as cooking fuel. However, as expected, children living in houses, which have sources of drinking water from hand pump/tap (public and

Table 2

Prevalence of Diarrhoea, Fever and Cough among Rural Children Born during the Three Years Preceding the Survey by Selected Housing Conditions and Parents' Socio-Economic and Housing Characteristics

Selected Housing Conditions and Parents' Socio-Economic Characteristics	Prevalence Rate							
	Diarrhoea		Fever		Cough		Total Children	
	AP	TN	AP	TN	AP	TN	AP	TN
Col. 1	Col. 2	Col. 3	Col. 4	Col. 5	Col. 6	Col. 7	Col. 8	Col. 9
Type of House								
Kachha	14.5	13.1	27.3	22.3	35.3	27.2	289	327
Semi-Pucca	15.2	13.9	26.0	23.2	35.3	25.9	269	259
Pucca	13.6	15.4	30.8	15.4	37.6	23.1	221	130
Density of Households@								
= 2	15.5	14.8	28.9	21.3	37.3	24.2	142	277
3 – 4	14.8	12.3	28.6	21.1	39.1	28.4	384	317
5 +	13.4	15.6	26.1	22.1	30.4	23.8	253	122
Separate Room used as a Kitchen								
No	12.9	13.8	26.2	24.2	34.9	29.9	542	298
Yes	18.1	13.9	31.6	19.4	38.4	23.2	237	418
Source of Drinking Water								
Well, River, Stream, etc.,	16.6	16.3	29.6	27.1	34.2	27.1	199	129
Hand Pump/Tap (Private/Public)	13.8	13.3	27.2	20.1	36.6	25.7	580	587
Type of Toilet Facility								
No Facility/Field/Bush	15.1	13.2	28.6	21.3	35.3	26.3	682	628
Flush Toilet	10.3	18.2	22.7	21.6	40.2	23.9	97	88
Main Source of Lighting								
Kerosene	16.3	22.7	22.1	26.5	32.7	34.3	208	181
Electricity	13.8	10.8	29.9	19.6	37.1	23.2	571	535
Type of Fuel for Cooking								
Biomass Fuels	14.8	13.2	27.8	21.9	35.4	26.6	698	608
Kerosene/LPG/Electricity, etc.	12.3	17.6	28.4	18.5	40.7	22.2	81	108

Contd...

Contd...

Education of Mother								
Illiterate	16.2	14.6	26.2	20.4	35.7	28.7	474	314
Literate,< Middle School completed	10.6	13.6	31.1	23.4	38.9	26.6	180	184
Middle School completed	16.0	12.0	40.0	22.2	34.0	24.8	50	117
High School completed and above	12.0	13.9	22.7	19.8	32.0	17.8	75	101
Work Status of Mother								
Not working	11.6	13.4	26.7	18.1	34.7	23.4	352	470
Working	16.9	14.6	28.8	27.6	37.0	30.9	427	246
Ownership of Durable Goods$								
= 3 Goods	16.8	15.8	26.4	22.2	36.1	28.8	382	361
4 – 7 Goods	14.1	10.0	29.0	20.0	35.0	20.8	283	260
8 + Goods	7.9	16.8	29.8	22.1	37.7	29.5	114	95
Total	14.5	13.8	27.9	21.4	35.9	26.0	779	716

Note: AP = Andhra Pradesh, TN = Tamil Nadu.

@ Calculated as total number of persons per room.

$ = Cumulative number of durable goods (11 items) available in the households which include Radio, Television, Bicycle, Motorcycle, Mattress, Pressure Cooker, Chair, Cot or Bed, Table, Clock or Watch and Fan.

private) and where there is electricity for lighting experience lower prevalence of diarrhoea. Across the socio-economic variables, differentials in the prevalence of diarrhoea among children are not that clear by their mothers' educational attainment. However, it appears to be somewhat more prevalent among children, whose mothers are working. Interestingly some sort of V–shape relationship is observed between the prevalence rate of diarrhoea among children and density of households as well as ownership of durable goods (col. 3 of Table 2).

Differentials in Fever

In rural Andhra Pradesh, contrary to the expectation, prevalence of fever is more among children living in pucca houses, less and moderately crowded houses, houses which have a separate room for kitchen and using electricity for lighting as against their counterparts. As expected, fever is less prevalent among children living in houses for which the source of drinking water is hand pump/tap (public and private) and households where there is some type of toilet facility. Unusually, the prevalence of fever among children is more among those whose mothers have some education (literate, <middle school complete and middle school completed), and households in which large number of durable household goods exist (col. 4 of Table 2).

In rural areas of Tamil Nadu, fever is less prevalent among children living in pucca houses, households having a separate room for kitchen, using drinking water from hand pump/tap (public and private), electricity for lighting and kerosene/gas/electricity as cooking fuel. Across the socio-economic variables, differentials in the prevalence of fever among children are not that clear by their mothers' educational attainment as well as ownership of durable household goods. However, the prevalence rate observed to be conspicuously higher among those children whose mothers are working than those who are not (col. 5 of Table 2).

Differentials in Cough

In the case of rural Andhra Pradesh, it is conspicuous to note that the differentials in the prevalence of cough across the housing environment factors are mostly contrary to what is normally expected. For instance, cough is comparatively more prevalent among children living in pucca houses, low and moderately crowded houses, houses that have a separate room for kitchen, households which are using water from hand pump/tap (private and public), having any type of toilet, using electricity for lighting and kerosene/gas/electricity as cooking fuel. The prevalence rate of cough among children neither differs much nor is consistent across the socio-economic characteristics of the parents under consideration, except across the work status of mothers; the prevalence rate of cough is higher among children born to working mothers than their counterparts (col. 6 of Table 2).

In rural areas of rural Tamil Nadu, as expected, cough is more prevalent among children living in kachha houses closely followed by semi-pucca houses, houses which do not have a separate room for kitchen and toilet facility, using kerosene for lighting and biomass fuels for cooking purpose. On the contrary, the prevalence of cough is comparatively more among children whose houses are moderate and less crowded, but did not vary much across the sources of drinking water. Across the socio-economic variables, there appears to be a consistent decrease in the prevalence rate of cough among children with an increase in the educational attainment of their mothers. Curiously, the prevalence of cough is more among those children whose mothers are participating in economic activities than those whose mothers did not. On the other hand, some sort of V-shape relationship is observed between ownership of durable household goods and the prevalence rate of cough among children (col. 7 of Table 2).

In the preliminary analysis, though an attempt is made to examine the differentials in the prevalence of diarrhoea, fever and cough across the selected socio-economic characteristics of the parents as well as housing conditions, viz., educational attainment and occupation of the father, caste background, size of owning cultivable land and irrigated land, and households own livestock or not, the results are found to be neither consistent nor that clear. Hence, they are not presented here, as well as not considered for further analysis.

Determinants of Child Morbidity Conditions

As noted earlier, in order to find out the principal determinants of child morbidity conditions under consideration, it is proposed here to adopt logistic regression analysis. This has been carried out with two models; in the first model, only housing conditions have been regressed with the morbid conditions and in the second model, selected socio-economic variables have also been included. For the purpose of these analyses, the dependant variables, viz., diarrhoea, fever and cough, have been considered as dichotomous, i.e., giving a value of '0' for absence of a particular disease and '1' for the presence, whereas most of the explanatory variables are self explanatory (see Tables 2 and 3).

Logistic Regression Analysis Results on Diarrhoea

Logistic regression results on diarrhoea of children in rural areas of Andhra Pradesh (col. 2 of Table 3) reveal that among the housing environment factors only two have turned out to be comparatively significant in showing their net effects; one on the expected direction and the other contrary to the expectation. For instance, children living in houses which do not have a separate room for kitchen are 0.6 times less likely to suffer with diarrhoea than those who live in houses that have such a facility ($p<0.05$); whereas, as expected, the odds of

Table 3

Results of Logistic Regression Analysis on Prevalence of Diarrhoea, Fever and Cough in Rural Andhra Pradesh and Tamil Nadu.

Explanatory Variables	*Diarrhoea*				*Fever*				*Cough*			
	Andhra Pradesh		*Tamil Nadu*		*Andhra Pradesh*		*Tamil Nadu*		*Andhra Pradesh*		*Tamil Nadu*	
	Model I	*Model II*	*Model I*	*Model II*	*Model I*	*Model II*	*Model I*	*Model II*	*Model I*	*Model II*	*Model I*	*Model I*
Col. 1	*Col. 2*	*Col. 3*	*Col. 4*	*Col. 5*	*Col. 6*	*Col. 7*	*Col. 8*	*Col. 9*	*Col. 10*	*Col. 11*	*Col. 12*	*Col. 13*
Type of House (Ref: Kachha)	1.000	1.000	1.000	1.000	1.000	1.000	1.000	1.000	1.000	1.000	1.000	1.000
Semi-Pucca	1.046	1.058	1.353	1.433	.862	.835	1.205	1.149	.920	.941	1.124	1.128
Pucca	.886	.971	1.595	1.618	1.092	1.131	.757	.703	.888	.938	1.053	.972
Density of Persons (Ref: = 2)	1.000	1.000	1.000	1.000	1.000	1.000	1.000	1.000	1.000	1.000	1.000	1.000
3 – 4	.918	.881	.905	.896	1.006	.957	.859	.891	1.113	1.058	1.033	1.057
5+	.841	.750	1.301	1.258	.946	.912	.781	.765	.774	.711	.659	.633
Separate Room used as Kitchen (Ref: Yes)	1.000	1.000	1.000	1.000	1.000	1.000	1.000	1.000	1.000	1.000	1.000	1.000
No	.599*	527**	.855	.826	0.818	.823	1.617+	1.514+	0.955	.906	1.616*	1.505+
Source of Drinking Water												
(Ref: Hand Pump/Tap – Private/Public)	1.000	1.000	1.000	1.000	1.000	1.000	1.000	1.000	1.000	1.000	1.000	1.000

Contd...

Contd...

Well, River, Stream, etc.,	1.118	1.227	1.267	1.036	1.226	1.202	1.638+	1.564+	0.911	.940	1.016	1.053
Type of Toilet Facility												
(Ref: Flush Toilet)	1.000	1.000	1.000	1.000	1.000	1.000	1.000	1.000	1.000	1.000	1.000	1.000
No Facility/Field/Bush	1.859+	1.050	.763	.740	1.785*	1.884*	.792	.845	0.819	.810	.901	.880
Main Source of Lighting												
(Ref: Electricity)	1.000	1.000	1.000	1.000	1.000	1.000	1.000	1.000	1.000	1.000	1.000	1.000
Kerosene	1.251	1.036	3.296***	3.064***	0.643*	.652*	1.520+	1.585+	0.884		1.719**	1.610*
Type of Fuel for Cooking												
(Ref: Kerosene/LPG/Electricity, etc.)	1.000	1.000	1.000	1.000	1.000	1.000	1.000	1.000	1.000	1.000	1.000	1.000
Biomass Fuels	1.065	0.799	.513+	.629	0.972	.937	1.287	1.382	0.871	.730	1.141	1.331
Education of Mother (Ref: Illiterate)												
Literate, < Middle School Completed	–	1.000	–	1.000		1.000	–	1.000	–	1.000	–	1.000
Middle School Completed		.734		1.019		1.532+		1.526+		1.123		1.033
High School Completed & above		1.208		.760		2.018*		1.311		.836		.889
		.818		.763		.810		1.162		.675		.536+

Contd...

Contd...

Work Status of Mother												
(Ref: Not working)	–	1.000	–	1.000		1.000		1.000		1.000		1.000
Working		1.291		.980	–	1.254	–	1.746**	–	1.183	–	1.433+
Ownership of durable goods												
(Ref: = 3 goods)	–	1.000	–	1.000	–	1.000	–	1.000	–	1.000	–	1.000
4 – 7 goods		.782		.735		.939		1.108		.821		.867
8 – 11 goods		*.331		1.058		1.052		1.839+		.808		2.246*

Note: +, *, ** and *** = Odds Ratios are Significant at .10, .05, .01 and .001 level. – = Variables not included in the Logistic Regression Models

diarrhoea are comparatively higher (1.86, $p<0.10$) for those children who live in households which do not have a toilet facility than those who live in households that have such a facility. Conspicuously, in the presence of socio-economic characteristics (col. 3 of Table 3), the net effect of not having a separate room for kitchen on the likelihood of diarrhoea among children has become more robust ($p<0.01$), but the effect of not having a toilet facility has become insignificant. On the other hand, the odds of experiencing diarrhoea by children are much lower when they belong to the households, which have large number of durable goods ($p<0.05$) than those who belong to the households of less number of durable goods.

In the case of rural Tamil Nadu, the odds of children who suffered with diarrhoea are observed to be significantly higher (3.30, $p<0.001$) if they belong to households where kerosene is used for lighting than their counterparts. On the contrary, such odds are comparatively lower (0.5 times, $p<0.10$) among those who belong to the households where biomass fuels (wood, cow-dung cakes, etc.) are used for cooking than those who live in households wherein kerosene/LPG/electricity is used for such purpose (col. 4 of Table 3). However, when the socio-economic factors included in the analysis (col. 5 of Table 3), while the net effect of using kerosene for lighting on the prevalence of diarrhoea remained significant, that too, in positive direction (odds ratio=3.06; $p<0.001$), the role of biomass fuels for cooking has become insignificant even at 10 per cent level.

Logistic Regression Analysis Results on Fever

In rural areas of Andhra Pradesh, logistic regression results on fever among children (col. 6 of Table 3) highlight that only type of toilet facility available in the house and main source of lighting found to have significant effect. However, while the effect of type of toilet facility on the risk of fever is in the expected direction, i.e., children living in houses which do not have toilet facility are significantly higher odds of suffering with fever (1.79, $p<0.05$) than those who live in houses that did not have such a facility; unexpectedly, children living in houses in which kerosene is used for lighting are reported to be at lower risk of suffering from fever (0.64 times, $p<0.05$) than those who live in houses wherein electricity is the main source of lighting (col. 6 of Table 3). It is pertinent to note that both of these variables maintained their relationships with fever even when the three socio-economic variables are included in the analysis (col. 7 of Table 3). Moreover, contrary to the expectation, the likelihood of children suffer from fever is observed to be significantly higher (1.5 and 2.0 times, $p<0.10$ and $p<0.05$, respectively) among those whose mothers have little education (literate, <middle school) as well as completed middle school education than their illiterate counterparts.

In the case of rural Tamil Nadu, a few housing environment variables have exhibited somewhat significant net effects, mostly in the expected direction, on the odds of children suffering with fever (col. 8 of Table 3). For instance, it is interesting to note that children living in houses which did not have a separate room for kitchen, use unprotected water (well, river, stream, etc.) for drinking and use kerosene for lighting are significantly at higher risk of suffering from fever (1.62, 1.64 and 1.52 times, respectively; $p<0.10$ in each case) than those who live in houses which have such facilities in a better way. It is conspicuous to note that even in the presence of socio-economic factors (col. 9 of Table 3) these net effects remained intact. Further, it is interesting to note that the likelihood of suffering with fever is higher among those children whose mothers are literates, but not completed middle school education (1.53 times, $p<0.10$), working for wages (1.75 times, $p<0.01$) and also among those who live in households that own large number of durable goods (1.84 times, $p<0.10$) as compared to those whose mothers are illiterates, not working and live in households that have less number of durable goods.

Logistic Regression Analysis Results on Cough

Logistic regression results on cough among children of rural Andhra Pradesh showed that the net effects of housing environment factors either in the absence or in the presence of socio-economic variables did not exhibit neither consistent nor significant net effects (col. 10 and 11 of Table 3).

In rural areas of Tamil Nadu, a few housing environment variables have exhibited significant net effects, mostly on the expected lines, on the likelihood of children suffering from cough. For instance, it is conspicuous to note that the probability suffering with cough is significantly higher among those children who live in houses that do not have a separate room for kitchen and use kerosene for lighting ($p<0.05$ and $p<0.01$, respectively) than those who live in houses that have a separate room for kitchen and use electricity for lighting (col. 12

of Table 3). Even when socio-economic factors are included in the analysis (col. 13 of Table 3), the direction of these net effects did not change, but the magnitude of effects has reduced a little. Another conspicuous fact noticed here is that education of mother, by and large, has exhibited negative effect on the likelihood of children suffering from cough and even turn out to be somewhat significant in the case of those children whose mothers completed high school and above level of education (0.54 times, $p<0.10$) as against their illiterate counterparts. On the other hand, unexpectedly, the odds of suffering with cough are noticed to be high among those children whose mothers are working (1.43 times, $p<0.10$) and among those who live in households that have large number of durable goods (2.25 times, $p<0.05$) than among those whose mothers are not working and who live in households that possess less number of durable goods.

Conclusions and Implications

Though it is difficult in finding out the relative effect of housing environment factors on child morbidity, as one has to consider other different sides of environment, based on the findings of the present paper some of the following conclusions have been drawn. With a few exceptions, by and large, the bi-variate results support most of the theoretical considerations with regard to the associations between housing environment factors and prevalence rates of diarrhoea, fever and cough among children. The results based on the logistic regression analysis highlight that in each state only 2-3 housing environmental factors turn out to have significant net effects, mostly in the expected direction, on the three child morbid conditions, except on cough in the case of rural Andhra Pradesh wherein none of them turn up significant. Moreover, the role of housing factors in determining child morbidity conditions is mostly logical and transparent, in majority of the cases, in rural Tamil Nadu than in Andhra Pradesh state. The pertinent conclusion one can see here is that the prevalence of the three child morbidity conditions is much higher among those children living in households in which kerosene is used for lighting, followed by households that did not have a separate room for kitchen and any toilet facility. Unprotected drinking water, by and large, showed positive effect on the prevalence of all three child morbid conditions under consideration, and also turned out to be somewhat significant in the case of cough in rural Tamil Nadu.

A few unexpected patterns were also noticed in some cases. For example, in rural Andhra Pradesh, children reported to be suffering less with diarrhoea and fever in spite of they live in houses that do not have a separate room for kitchen and wherein kerosene is used for lighting. Another unexpected pattern is that the prevalence of diarrhoea among children of rural Tamil Nadu is comparatively lower for those who live in houses which use biomass fuels for cooking. While such households might mostly belong to lower socio-economic status, especially mothers may be with little education, less reporting of sickness of children by these women is likely which is widely reported from other studies; of course, a detail examination is also indispensable. The former contention is further supported by the findings of this study where mothers completed middle school education in rural Andhra Pradesh and mothers who are literate, but not completed middle education in rural Tamil Nadu have reported fever for their children on a higher side. Likewise, the ownership of durable goods, proxy for economic status, has also demonstrated a positive net effect on fever and cough in rural Tamil Nadu. Further, among the socio-economic factors, work participation of mothers invariably, except in one case, showed a positive net effect on the prevalence of child morbidity conditions considered here, in both the states, and the results even turned out to be significant in the case of fever as well as cough in rural Tamil Nadu. Among working women less time at their disposal for child care, not able to provide the preventive care with the same intensity to their children as non-working women, and at times even forgo breastfeeding (Sivakami, 2001).

In the light of the findings and conclusions drawn from the study, the following policy suggestions are put forth so as to see that the common illnesses that are caused by housing environment factors would not occur among rural children. First of all, people in general and parents in particular, have to be informed about the detrimental effects of housing conditions on child health and the need for improvement of the housing environment, especially, using electricity for lighting, keeping up a separate room or space for kitchen outside the living room and taking protected drinking water Providing subsidies to construct houses with better conditions, providing protected drinking water through taps, and toilet facilities at villages by the local self-government would enhance the housing environment and thereby, child health. Improving socio-economic conditions of the people in general and that of women in particular is also essential so as to provide better facilities to their children and also visiting health facilities at the earliest for preventive as well as for curative services when children fell sick.

References

Agarwal, D.K. and G.P. Katiyar (1981): "Influence of Environmental Factors on Under Five Morbidity", *Indian Pediatrics*, 18(8): 545-550.

Boerma, J.T., A.E. Sommerfelt and O.R. Shea (1991): "Childhood Morbidity and Treatment Patterns", *DHS Comparative Studies,* No. 4. Columbia, Maryland: Institute for Resource Development.

Chojnacka, H. and O. Adegbola (1984): "The Determinants of Infant and Child Morbidity in Lagos, Nigeria", *Social Science and Medicine* 19(8): 799-810.

International Institute for Population Sciences (IIPS) and ORC Macro (2000) "National Family Health Survey (NFHS-2), India 1998-99: Andhra Pradesh", Mumbai: IIPS.

– (2001): "National Family Health Survey (NFHS-2), India 1998-99: Tamil Nadu", Mumbai: IIPS.

Majumdar, A.K. and S.M.S. Islam (1993): "Socio-economic and Environmental Determinants of Child Survival in Bangladesh", *Journal of Biosocial Science,* 25(2): 311-318.

Mishra, V. and R.D. Retherford (1997): "Cooking Smoke Increases the Risk of Acute Respiratory Infection in Children" *NFHS Bulletin,* 8: 1-4.

Muttalib, M.A. (1990): *Child Development,* New Delhi: Sterling Publishers Private Ltd.

Park, J.E., and K. Park (2000): *Textbook of Preventive and Social Medicine,* Jabalpur: Banarsidas Bhanot Publishers.

Retherford, R.D. and M.K. Choe (1993): *Statistical Models for Causal Analysis.* New York: John Wiley.

Sethapongkul, S. (1992): "The Effects of Socio-economic Factors on Adult Morbidity in Thailand", *Health Transition Review,* 2(2): 228-229.

Sivakami, M. (2001): "Mother's Work Participation and Child Health: A Study in Poor Localities in Tamil Nadu, India". Unpublished Ph. D. Dissertation. Coimbatore: Department of Population Studies, Bharathiar University.

Tekce, B. and F.C. Shorter (1984): "Determinants of Child Mortality: A Study of Squatter Settlements in Jordan", *Population and Development Review,* 10(Supplement): 257-279.

Timaeus, I.M. and L. Lush (1995): "Intra-urban Differentials in Child Health", *Health Transition Review,* 5(2): 163-190.

United Nations Children's Fund (UNICEF), (1991): *The State of the World's Children, 1991,* New York: Oxford University Press.

Vidya, P. (1997): "Effect of Living Conditions on Child Morbidity and Treatment Pattern in Rural Andhra Pradesh and Tamil Nadu". Unpublished M. Phil. Dissertation, Coimbatore: Department of Population Studies, Bharathiar University.

Visaria, L. (1988): "Levels, Trends and Infant Mortality in India", in A. K. Jain and P. Visaria (eds.), *Infant Mortality in India,* New Delhi: Sage Publications, pp. 67-126.

15 Effect of Household Environment on Women's Health in North-East India

Avishek Hazra, Subhra Datta and Mohua Guha

The growing propensity of environmental pollution brings a great threat to people of this planet. Environment, which comprises physical, biological as well as social factors, is a crucial determinant of human health. Women must be the ambassadresses of environmental issues, as the environment starts from the home. Women and environment are inseparable. Each woman can contribute significantly to protect her environment and it is a bounden duty too. Most women spend about 90 per cent of their time indoors in their homes or other buildings. So, for people who spend most of their time within the household, the health risks of indoor air pollution are greater than those outdoors. Gases are the main cause of indoor air problems in homes. Their sources may include many things like (i) oil, gas, coal, wood, kerosene and tobacco products, (ii) materials used to build home, (iii) home furnishings, such as cabinets made of certain pressed wood products, (iv) cleaning, pest control, painting, and personal care products, etc.

It is a common practice among villages to use wood, cow dung, crop residues for cooking and these are extensively used in a large number of households even today. But they are unaware of the consequences and damage caused by indoor air pollution. Studies show that these fuels emit a large number of pollutants in high quantities. They are so concentrated that a ton of particulates from household stoves could lead to more than 500 times the human exposure than a ton of particulates from a coal-fired power plant (CSE, 1985). This conventional and traditional practice is hard to change by. Most rural Indian homes use traditional 'chullhas', which are fuel-inefficient. Having poor air flow or ventilation, in combination with heating and cooling systems can also cause indoor air problems.

Review of Literature

Various researches show that indoor air pollution has a significant impact on the health of all household members, particularly that of women and children. Polluted indoor and outdoor air, contaminated water, lack of adequate sanitation, toxic hazards, disease vectors, ultra-violet radiation, and degraded ecosystems are all important environmental risk factors for children and their mothers as well.

Health effects from indoor air pollution may start right away or occur years later. These things can happen after being exposed once or many times. Health effects may show up years later, or only after long periods of being exposed. These health effects include heart disease, cancer, breathing and lung problems. Indoor air pollution is responsible for 2.7 per cent of the global burden of disease, 2.5 per cent in males and 2.8 per cent in females (WHO, 2002). Recently, WHO estimated the global distribution of burden of disease by risk factors and it showed that the environmental risk factors like unsafe water, sanitation and hygiene, indoor smoke from solid fuels together accounted for around 15 per cent in developing countries.

Prolonged exposure to high level of pollutants caused mostly by biomass fuels has a very adverse impact on the health of women who spent three to five hours everyday cooking with biomass fuels. The effects could include eye problems, respiratory problems, chronic bronchitis, lung cancer, Cor-Pulmonate (a heart disease) and adverse reproductive outcomes.

Research studies shows that use of biomass fuels for cooking, substantially increases the risk of tuberculosis in India for persons aged 30 and above (Mishra et al., 1997). In another study Salmond (1999), found significant association between asthma and deprived areas. Rona (1999) found a disproportionate increase of asthma in lower social classes and marginal relation to family size. An only child had a higher prevalence of asthma. Thus, rapid deterioration in the quality of environment has overburdened the health system in many countries. Air and water pollution are directly responsible for many of the health problems and it is evident that the household environment is a major contributor to this burden. The World Bank in 1993 estimated total burden of disease attributable to the problems related to the household environment including overcrowding, lack of facilities such as sanitation, garbage disposal, indoor air pollution and vector-breeding grounds, etc. The study further concluded that 30 per cent of the burden of disease could be avoided by improvements in the household environment, of which 20 per cent can be achieved through modest interventions alone. The estimates of the WHO (1997) indicate that although virtually all cases have an environmental cause, 90 per cent of the diarrhoeal diseases could be avoided through 'feasible environmental interventions'. In all, environmental factors account for about 23 per cent of the global burden of disease.

In view of above discussion, the present paper examines the effect of household environmental factors on the health outcomes of women in the north-eastern states in India. To capture the health perspective, we have taken on two air-borne morbidities: asthma and tuberculosis that were available in the used dataset.

Data and Methods

The data which has been used for the present study is extracted from National Family Health Survey (NFHS-II) conducted in 1998-99. The present analysis is based on women aged 15 and above living in the sample households of the north-eastern states of India and have considered morbidities caused mainly from air pollution.

The household environment has been measured using data for basic household amenities from the household information. The variables included for measuring household environment are: type of house, type of fuel used for cooking, room facility, sources of lighting and existence of separate room for kitchen. To understand the relation between housing condition and morbidity, simple frequency distribution and bi-variate analysis has been performed. Relative risks and odds ratios have been calculated by using logit regression that explains the influence of Asthma and TB on household environment conditions. The analysis is carried out for the seven north-eastern states of India from the NFHS-II data.

Results and Discussion

Table 1 gives the socio-economic and demographic profile of the population in the north-eastern states of India. Among the seven neighbouring states, four states, viz., Tripura, Assam, Manipur and Sikkim are Hindu dominated. On the other hand, Mizoram, Nagaland and

Table 1

Socio-Economic and Household Environment Profile of North-Eastern States: NFHS-II Data

Characteristics		*Assam*	*Mani-pur*	*Megha-laya*	*Mizo-ram*	*Naga-land*	*Sikkim*	*Tri-pura*	*North East*
Sex	Male	49.5	47.7	48.2	47.8	49.4	49.6	48.9	48.9
	Female	50.5	52.3	51.8	52.2	50.6	50.4	51.1	51.1
Religion	Hindu	66.7	50.9	10.0	1.9	11.4	61.2	87.6	50.9
	Muslim	29.5	5.2	3.6	0.3	7.1	1.2	9.5	14.1
	Christian	2.1	32.6	70.1	96.3	80.1	5.0	1.6	27.5
	Others	1.6	11.2	15.6	1.5	1.4	32.7	1.3	7.5
Caste	SC	63.3	51.4	4.8	0.4	12.2	72.8	78.3	49.4
	ST	14.7	34.7	89.8	99.1	82.6	27.1	13.5	38.3
	Others	21.9	13.9	5.4	0.5	5.2	0.0	8.3	12.3
Education	No Education	32.7	20.3	30.0	8.2	26.2	32.3	22.4	26.6
	Primary	18.2	16.8	30.9	27.8	20.9	24.9	28.7	22.1
	Secondary	35.3	40.9	31.4	51.8	43.2	32.9	38.3	38.1
	Higher	13.8	22.0	7.7	12.2	9.7	9.9	10.6	13.3
SLI	Low	32.2	26.1	35.4	8.5	24.6	11.1	28.9	26.0
	Medium	48.3	56.8	53.8	65.5	60.2	67.8	59.6	55.9
	High	19.5	17.1	10.8	25.9	15.2	21.1	11.5	18.1
Place of Residence	Rural	78.0	63.6	76.4	41.1	79.8	87.8	84.2	73.8
	Urban	22.0	36.4	23.6	58.9	20.2	12.2	15.8	26.2
Type of House	Pucca	17.5	8.0	17.7	20.7	22.2	48.8	7.3	19.0
	Semi-Pucca	41.2	18.0	35.2	54.5	18.8	42.3	6.2	33.6
	Kachha	41.4	74.0	47.1	24.8	59.0	8.9	86.5	47.3
Separate Kitchen	Yes	85.1	76.3	83.1	49.1	90.8	64.1	76.2	77.1
	No	14.9	23.7	16.9	50.9	9.2	35.9	23.8	22.9
Type of fuel used	Modern	19.3	29.4	9.0	45.5	13.4	19.9	13.5	21.7
	Traditional*	80.7	70.6	91.0	54.5	86.6	90.1	86.5	78.3
No. of members per room	Up to 3	77.0	77.6	79.8	57.2	87.0	79.6	64.5	74.9
	3 and above	23.0	22.4	20.2	42.8	13.0	20.4	35.5	25.1

* Includes wood, crop residues, dung cakes, coal/coke/lignite, charcoal and kerosene.

Meghalaya are Christian dominated states having 96 per cent, 80 per cent and 70 per cent of Christians respectively. In these three states, percentages of Scheduled Tribes (STs) population are found to be high. Almost all population of Mizoram belongs to this specific group. The percentages for the same in Meghalaya and Nagaland are 90 per cent and 83 per cent respectively. In the remaining four states, Scheduled Castes (SC) are major ethnic group. Illiteracy is still prevailing in these states; except the state of Mizoram, where 92 per cent people have basic education. Majority of the population are with medium standard of living and except Mizoram, all these states are constituted of rural population. Most of the people live in kachha houses and majority do not have any separate room for kitchen. In North-East India, around 78 per cent of population depends upon traditional means of cooking which includes wood, crop residues, dung cakes, coal/coke/lignite, charcoal and kerosene. The state-wise percentages for the same lie between the small range of around 70 per cent to around 90 per cent. It is seen from the result that most of the households have adequate room facility (less than 3 persons per room). About 75 per cent of the households in the north-eastern states have the facility of maximum three people living in one room. There is no variation in the percentages of this variable among the states.

The prevalence of asthma among the women aged 15 and above according to several socio-economic, environmental and personal habits related variables is given in Table 2. It is evident from Table 2 that prevalence of asthma among the women is highest in Sikkim followed by Nagaland, Meghalaya and Tripura. Taking North-East as a whole, the prevalence of asthma is around four. The prevalence is 3.9 for the women belonging to rural areas and the same for urban women is 2.5 and similar pattern follows in almost all the north-eastern states, i.e., women living in rural areas are at more risk from suffering the disease compared to their urban counterparts. Quite obviously with the increase in age, the prevalence also increases. Women belonging to low Standard of Living Index (SLI) category suffer more from this respiratory disease asthma than the women belonging to high or medium SLI.

It is seen that except few states like Tripura, Nagaland and Mizoram, in most of the north-eastern states, women who are using traditional fuel, are more susceptible to this disease compare to those using modern fuels. It may be argued that, in Tripura, some other environmental factors such as dampness in the weather which is very much influential in spreading of asthma virus 'azmone' and also excessive rubber plantation in the state increase the risk of asthma.

Table 2

Prevalence of Asthma among Women Aged 15 and above According to Various Household Environment and Background Characteristics in North-Eastern States: NFHS-II Data

Asthma		*Assam*	*Manipur*	*Meghalaya*	*Mizoram*	*Nagaland*	*Sikkim*	*Tripura*	*North East*
Type of fuel used	Modern	1.3	1.8	1.3	2.8	6.9	5.8	7.7	2.8
	Traditional*	2.5	2.4	4.8	2.1	5.5	8.7	3.9	3.7
Type of House	Pucca	1.7	0.4	1.9	1.8	3.5	5.5	6.6	3.0
	Semi-pucca	2.3	2.8	3.5	2.7	7.9	11.0	7.8	4.0
	Kachha	2.4	2.3	6.1	2.2	5.8	8.5	4.0	3.4
Smoke Regularly	No	2.1	1.7	4.0	1.5	5.7	6.8	4.1	3.1
	Yes	19.0	15.4	N.A	10.2	11.1	22.1	20.0	14.7
Chew Tobacco	No	2.0	1.9	4.5	2.6	6.0	7.1	4.5	3.4
	Yes	3.3	3.3	4.4	2.3	4.3	12.0	3.7	3.9
No. of members per room	<=2	2.1	2.7	4.4	2.3	5.6	8.0	4.4	3.6
	3-5	2.8	1.1	4.4	2.7	6.6	8.0	4.6	3.5
	6+	1.4	2.4	5.6	1.4	N.A	8.7	3.5	3.2
Separate room for Kitchen	Yes	2.1	2.5	4.1	1.9	5.6	7.1	4.7	4.4
	No	3.4	1.5	6.0	2.9	7.3	9.7	3.7	3.3
Source of lighting	Others	1.7	2.1	3.5	2.5	4.9	7.4	4.5	3.4
	Kerosene	2.6	2.6	5.5	1.9	7.2	10.6	4.3	3.7
Age group	15-29	1.3	1.3	3.4	0.5	4.3	4.7	2.7	2.1
	30-59	2.8	2.4	4.3	3.0	6.7	9.6	4.7	4.0
	60+	7.2	7.0	15.8	12.4	17.0	28.5	11.1	11.4
SLI	Low	2.6	2.6	6.2	0.6	8.6	10.9	3.7	3.8
	Medium	2.4	2.3	3.8	2.5	4.5	8.8	4.2	3.7
	High	1.3	1.8	2.7	2.7	5.8	4.4	7.4	2.7
Place of Residence	Urban	1.5	2.3	2.0	2.3	4.6	4.9	5.6	2.5
	Rural	2.5.	2.2	5.2	2.7	6.0	8.5	4.2	3.9
Overall		2.3	2.2	4.4	2.4	5.7	8.1	4.4	3.5

* Includes wood, crop residues, dung cakes, coal/coke/lignite, charcoal and kerosene.

The prevalence of asthma is more among the women who are staying at kachha or semi-pucca houses compared to those of staying at pucca houses. Separate room for kitchen and source of lighting in the house also influences the incidence asthma among women. Those women using kerosene as the source of lighting are more exposed to suffer from this disease compared to those using others means as a source of lighting. Personal habits of smoking and chewing tobacco evidently increase the risk of suffering from this disease.

Table 3

Prevalence of Tuberculosis among Women Aged 15 and Above According to Various Household Environment and Background Characteristics in North-Eastern States: NFHS-II Data

Tuberculosis		*Assam*	*Manipur*	*Meghalaya*	*Mizoram*	*Nagaland*	*Sikkim*	*Tripura*	*North East*
Type of fuel used	Modern	0.5	0.6	0.6	1.5	2.9	1.4	N.A	0.9
	Traditional*	0.7	1.2	1.8	1.1	1.8	1.2	1.3	1.1
Type of House	Pucca	0.5	0.4	1.6	0.9	1.4	1.3	0.7	0.9
	Semi-Pucca	0.8	0.8	0.5	1.2	2.5	1.4	N.A	1.0
	Kachha	0.5	1.1	2.6	1.8	2.0	0.5	1.2	1.2
Smoke Regularly	No	0.6	0.8	1.4	1.3	1.9	1.2	0.7	0.9
	Yes	N.A	1.9	N.A	1.5	N.A	4.4	N.A	2.0
Chew Tobacco	No	0.5	1.0	2.0	1.5	2.2	1.3	1.0	1.0
	Yes	1.2	0.8	0.8	1.2	0.9	1.2	3.7	1.2
No. of members per room	<=2	0.7	1.0	2.0	1.3	2.0	1.4	0.7	1.1
	3-5	0.6	1.0	0.7	1.4	2.1	1.2	1.3	1.0
	6+	N.A	1.2	2.8	0.7	N.A	N.A	3.5	1.0
Separate room for Kitchen	Yes	0.7	1.1	1.6	1.0	1.9	0.7	0.6	0.9
	No	0.5	0.7	2.2	1.6	2.7	2.2	2.8	1.5
Source of lighting	Others	0.5	0.9	0.6	1.2	2.5	1.4	0.9	1.0
	Kerosene	0.7	1.5	2.9	2.4	1.1	0.5	1.5	1.1
Age group	15-29	0.4	0.7	1.5	1.0	1.7	1.1	0.4	0.7
	30-59	0.7	0.5	2.0	1.7	2.3	1.5	1.8	1.3
	60+	2.5	1.2	1.1	1.3	2.1	0.8	1.4	2.0
SLI	Low	0.9	1.4	2.0	2.4	2.6	1.7	2.2	1.4
	Medium	0.6	1.0	1.7	1.0	1.6	1.1	0.8	1.0
	High	0.4	0.6	1.1	1.8	2.1	1.6	NA	0.9
Place of Residence	Urban	0.6	0.8	0.7	1.4	1.2	2.3	NA	0.9
	Rural	0.7	1.1	2.0	1.1	2.2	1.1	1.3	1.1
Overall		0.7	1.0	1.7	1.3	2.0	1.3	1.1	1.1

* Includes wood, crop residues, dung cakes, coal/coke/lignite, charcoal and kerosene.

Table 3 depicts the prevalence of tuberculosis among the women aged 15 and above according to several socio-economic, household environmental and personal habits related variables. It is visible from the table that though the prevalence of tuberculosis is not that much, the variables related to household environment do have influence on this disease too.

The odds ratios for these two diseases, asthma and tuberculosis among the women in these north-eastern states by selected indicators related to household environment are given in Table 4. Results from Table 4 clearly supports that household environmental factors have a very negative role on women's health. The risk of suffering from asthma is 1.3 times more among women who use traditional means of fuel for cooking than those of modern or clean fuel user. It seems from the result that adequate room facility is not behaving in the expected direction, but it should be noted here that less sample size may not depict the true picture. Women who use a separate room as a kitchen are at 1.3 times lower risk than those of not having the facility.

The calculated odds ratio indicate that, in North-East India, women who use unclean, traditional fuel are 1.2 times more likely to suffer from tuberculosis than those of using modern means of clean fuels. In Meghalaya, women in the former group are at 2.9 times higher risk than the later group. Women residing in kachha houses are at 1.2 times higher risk than the women who are residing in pucca houses. Separate room for kitchen is also playing substantial role in most of these states. In Tripura, women who are not having any separate room for kitchen are five times more likely to suffer from tuberculosis and for north-east as a whole, the odds for the same is 1.6. Use of kerosene increases the risk of suffering from both the diseases.

Conclusion

The analysis clearly illustrates the morbidity situations may be attributable to household air pollution. It tells how the factors related to household environment influence the prevalence of diseases and thus effects the health of women. Women residing in rural areas and of low standard of living are at more risk of suffering from the respiratory morbidities of asthma and tuberculosis. Use of unclean fuel increases the likelihood of suffering from these types of morbidities. Women residing in semi-pucca and kachha houses are more likely to suffer from these respiratory disorders. Personal habits like chewing tobacco

Table 4

Odds Ratio of Asthma (As) and Tuberculosis (TB) among Women in the North-Eastern States by Selected Indicators Related to Household Environment, NFHS-II

	Assam		Manipur		Meghalaya		Mizoram		Nagaland		Sikkim		Tripura		North East	
	As	*TB*	*As*	*TB*	*As*	*TB*	*As*	*TB*	*As*	*TB*	*As*	*TB*	*As*	*TB*	*As*	*TB*
Type of fuel used																
Modern	–	–	–	–	–	–	–	–	–	–	–	–	–	–	–	–
Traditional*	1.90	1.28	1.36	2.12	3.89	2.86	0.75	0.70	0.79	0.62	1.54	0.88	0.49	–	1.34	1.21
Type of House																
Pucca	–	–	–	–	–	–	–	–	–	–	–	–	–	–	–	–
Semi-Pucca	1.40	1.56	6.47	1.71	1.82	0.32	1.50	1.35	2.36	1.80	2.14	1.10	1.20	0.06	1.33	1.05
Kachha	1.45	0.96	5.14	2.52	3.32	1.60	1.24	2.03	1.69	1.42	1.60	0.42	0.59	1.87	1.13	1.24
No. of members per room																
<=2	–	–	–	–	–	–	–	–	–	–	–	–	–	–	–	–
3-5	1.34	0.95	0.41	1.05	0.99	0.34	1.17	1.06	1.19	1.05	0.98	0.85	1.05	1.89	0.97	0.93
6 & above	0.67	0.05	0.88	1.23	1.28	1.40	0.58	0.51	0.01	0.01	1.09	0.03	0.80	5.30	0.89	0.89
Separate room for kitchen																
Yes	–	–	–	–	–	–	–	–	–	–	–	–	–	–	–	–
No	1.69	0.66	0.59	0.68	1.47	1.44	1.51	1.70	1.33	1.46	1.42	3.32	0.78	5.13	1.34	1.63
Source of lighting																
Others	–	–	–	–	–	–	–	–	–	–	–	–	–	–	–	–
Kerosene	1.61	1.47	1.24	1.71	1.62	5.09	0.76	2.01	1.51	0.42	1.48	0.35	0.96	1.61	1.11	1.14

* Includes wood, crop residues, dung cakes, coal/coke/lignite, charcoal and kerosene.

and smoking give rise to the risk of tuberculosis as well as asthma. Women belonging to households having the facility of separate kitchen are at lesser risk to get morbid and source of lighting that increases the level of indoor pollutants is also affecting the health of women.

Thus, attention must be given to educate people about the harmful impact of air pollution taking place in household environment. Promotion of smokeless 'chullha' may reduce health hazards, particularly among women who are using traditional fuel like firewoods, cow dung, and coal for cooking purpose and may bring other benefits like fuel wood conservation and ecological improvement. Less-polluting biomass fuels and other cleaner fuels like LPG (liquefied petroleum gas) could also decrease the risk of suffering from the respiratory diseases in a better ventilated housing environment. Increasing the role of women in decision-making concerning household matters such as the choice of fuel, stove, etc., will directly benefit the health of the women as well as the health of the family.

References

Bruce, N., R., Perez-Padilla, R., Albalak (2000): "The Health Effects of Indoor Air Pollution in Developing Countries", *WHO Bulletin* No.78: 1080-1092.

Centre for Science and Environment (CSE) (1985): *State of India's Environment: The Second Citizens' Report,* CSE, New Delhi.

International Institute for Population Sciences and ORC Macro. *National Family Health Survey (NFHS-2), India 1998-99*, Mumbai: International Institute for Population Sciences.

Lewis, S.A., S.T. Weiss, T.A. Platts-Mills, H. Burge, D.R. Gold (2002): "The Role of Indoor Allergen Sensitization and Exposure in Causing Morbidity in Women with Asthma", *American Journal of Respiratory Critical Care Medicine,* 156:23-27.

Mishra, V. and Rutherford Robert D. (1997): "Effects of Cooking Smoke on Prevalence of T.B. in India", *Population Series Paper No. 92*, October 1997.

Mishra, V.K. (2003): "Effect of Indoor Air Pollution from Biomass combustion on Prevalence of Asthma in the elderly", *Environmental Health Perspectives,* 111: 71-77.

Mishra, V.K., R.D. Retherford, and K.R. Smith (1999): "Biomass Cooking Fuels and Prevalence of Tuberculosis in India," *International Journal of Infectious Diseases, 3*: 119-129.

Nusselder, W.J. (2000): "Smoking and the Compression of Morbidity", *Journal of Epidemiology and Community Health,* Vol. 54, pp. 566-574.

Rona, R.J. (1999): "Association between Asthma and Family Size Between 1977 and 1994", *Journal of Epidemiology and Community Health,* Vol. 53, pp. 15-19.

Salmond, C. (1999): Asthma Prevalence and Deprivation: A Small Area Analysis, *Journal of Epidemiology and Community Health,* Vol. 53, pp. 476-480.

Smith, K.R., J. Zhang, R. Uma, Y. Ma, V.N. Kishore and K. Lata (1999): "Carbon Monoxide from Cooking Stoves in Developing Countries and Exposure Potentials, Chemosphere", *Global Change Science,* pp. 367-75.

World Bank. www.wbln.org

World Development Report (1993): *World Development Report. 1993: Investing in Health,* Oxford University Press, New York.

World Health Organization (1972): *Health Hazards of the Human Environment,* WHO, Geneva.

– (1997): *Health and Environment in Sustainable Development,* WHO, Geneva.

– (2002): *World Health Report: Reducing Risks, Promoting Healthy Life,* WHO, Geneva.

– www.who.int/en

Index